Building Responsive Library Collections with the Getting It System Toolkit

By Tim Bowersox, Cyril Oberlander,
Kate Pitcher, & Mark Sullivan

Contributing Authors:

Kerri Goergen-Doll, Oregon State University

Eric Joslin, Washington University in St. Louis

Ryan Litsey, Texas Tech University

Micquel Little, Monroe Community College, formerly at St. John Fisher College

Katherine Mason, Central Michigan University, formerly at Old Dominion University

Kate Ross, St. John Fisher College

Susanna Van Sant, Tompkins Cortland Community College

978-0-9897226-8-1 (Print)

978-0-9897226-9-8 (eBook)

IDS Project Press

Milne Library

1 College Circle

Geneseo, NY 14454

Contents

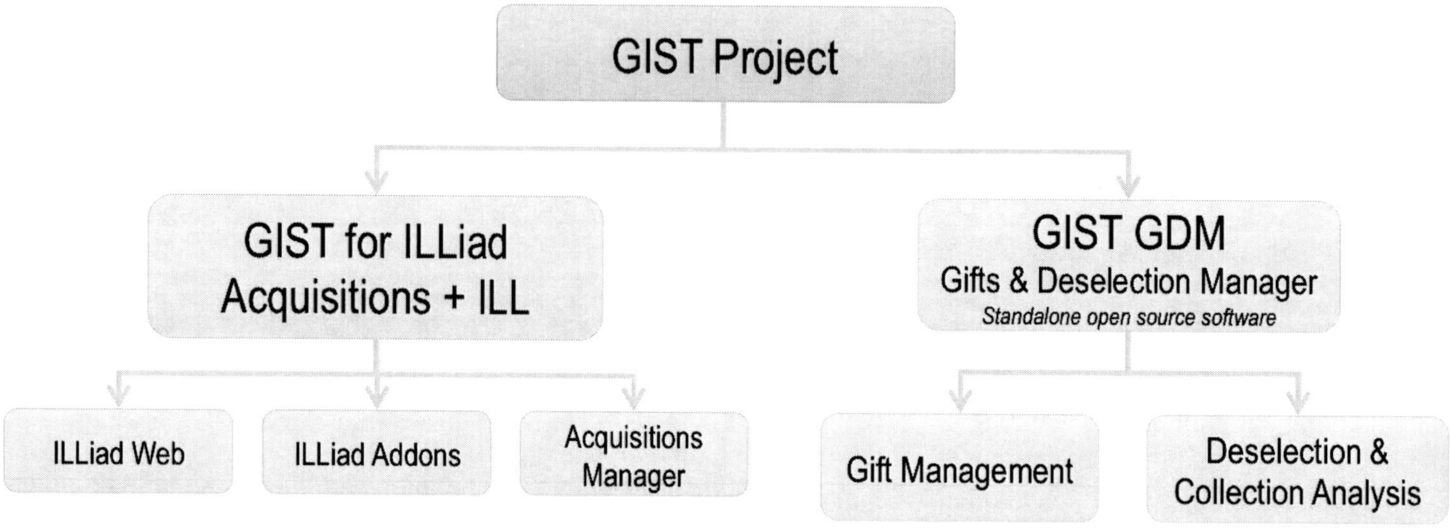
GIST Project
GIST for ILLiad
Acquisitions + ILL
GIST GDM
Gifts & Deselection Manager
Standalone open source software
ILLiad Web
ILLiad Addons
Acquisitions
Manager
Gift Management
Deselection &
Collection Analysis

Introduction

By Cyril Oberlander, Humboldt State University, formerly at SUNY Geneseo

GIST data automation and workflow mission critical for libraries' future

Fundamental to the creation of the Getting It System Toolkit (GIST) was an understanding that by automating processes that support better and more informed decisions, employees can be free to undertake higher-value tasks. The key automation features in GIST are a combination of gathering and sorting information, a recommendation based on local preferences, and lastly, a human's decision. This decision making framework allows libraries to streamline key functions so that they can expand services or offer new ones.

A mission-critical system is one that is so intertwined with the operation of an organization that the organization can barely function without it. It unites functions such as order, inventory, account management, and billing; accesses shared data repositories to eliminate redundant data entry; and streamlines access to data, making workflows more efficient. How are these corporate systems relevant to the library? They unlock talent and time. The talent and time of the academic library staff are exactly the resources we need to turn loose to innovate, create, and push out new services, and they are essential to the transformation of higher education and the learning environment.

Many college and university administrators agree and are looking for ways to engage in library transformation. This was most evident in the 2011 report, "Redefining the Academic Library, Managing the Migration to Digital Information Services, by the Advisory Board Company" (Education Advisory Board, 2011).That report showed that the demands on the library "require that top administrators, faculty, and students all work together to reach a consensus about how the library can best support the academic mission" (p. 14). The report asked whether libraries should pursue a variety of onsite and web services, instructional initiatives, and research support to meet the changing needs of their educational institutions. One of the many identified strategies was a need to shift to a demand-driven acquisitions model of collections.

Libraries are transforming and creating key roles and services that will shape their future. However, in order to pilot, develop, and implement these roles and services, every library needs staff time. Just as mission-critical systems in the corporate world deliver the freedom to innovate, they can be used to achieve the same for the library.

Fundamental to workflow transformation at the Milne Library at the State University of New York, Geneseo College was recognizing the central role of data gathering in library workflows and decision making, and the surprisingly similar data uses given various scenarios.' Recognizing that data and automation are critical to libraries, GIST was developed as two decision-making structures: one for purchasing and borrowing a book using ILLiad; and the other, Gift and Deselection Manager (GDM), a standalone system for evaluating and processing gift books and evaluating collections.

Unifying borrow/buy decision-making using The Getting It System Toolkit for ILLiad

The construction of the Getting It System Toolkit (GIST, GISTLibrary.org) started by rethinking the relationship between acquisitions and interlibrary loan (ILL). The process of purchasing and of borrowing a book are similar and rely on similar data. We streamlined the workflow of both the requester and library staff by gathering data that each found useful to process (requester submitting the request, library staff processing the request) and making the data available on demand and in context to their workflow. By adapting and extending ILLiad, a request management system made by Atlas Systems, Inc., GIST empowered users to make better decisions about their requests, and united the work and strategies shared across the two library units. Building on another platform was a strategy that also freed the GIST team from developing a purchasing/borrowing system from scratch, as we later had to do with the GIST Gift and Deselection Manager.

Table 1. Examples of contextual data used by library staff.

Purchasing a Book	**Borrowing a Book**	**Evaluating a Gift Book**	**Evaluating Weeding a Book**
Requester Holdings (local, consortia, state, etc.) Collecting area/scope (Library of Congress range) Cost	Requester Holdings (local, consortia, state, etc.) Collecting area/scope (Library of Congress range) Cost (free or fee)	Donator Holdings (local, consortia, state, free online etc.) Collecting area/scope (Library of Congress range) Cost	Use data Holdings (local, consortia, state, free online etc.) Collecting area/scope (Library of Congress range) Cost

The GIST Team at SUNY Geneseo developed GIST for ILLiad because we saw a need to reshape processes and empower both the end-user and library staff. Visits to many libraries that use ILLiad showed us that libraries often design workflows to match existing, segmented processes and policies. Instead of doing this, the development team applied the principles of seeking simplicity and sense-making to shape a new and more supportive experience. The goal was to apply automation with flexibility and the ability to customize. How these three factors work together is what matters to the user. The resulting sense-making workflow successfully converges two siloed processes; in fact, it works better if the strategies of acquiring and borrowing work together.

The sense-making workflow begins with two key questions. The first is how would you like to access an item? Users are presented that question in very separate contexts and web pages without GIST. With GIST, the request experience is curated to ask the question and seamlessly combine library processes; when a patron requests an item that is not in the library collection, the question is, "Should we buy it or borrow it?" There has never been a library system that handled that question well. Before GIST, a patron made an interlibrary loan request, and possibly the ILL staff sent an email to acquisitions staff about purchasing the item.

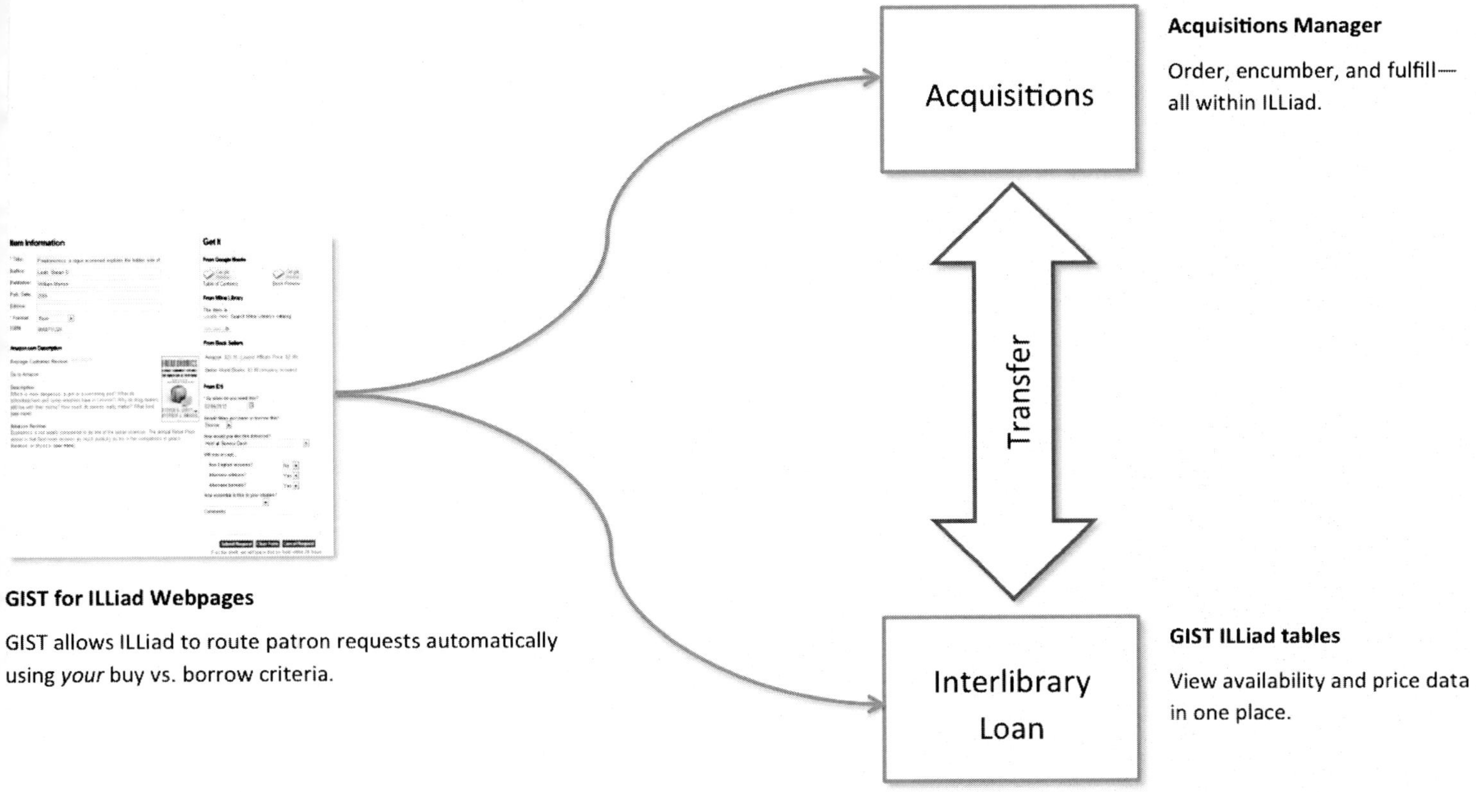

Figure 1. A broad overview of the GIST for ILLiad workflow.

GIST augments ILLiad's web request interface to add acquisition functionality. The request form behaves as both a request to purchase and a request to borrow, and returns all the data and web services that the library staff need to choose the best option. Useful data are made available to users and staff—they can see the book cover and reviews to evaluate the item, options for purchasing the book, full-text availability for immediate access, and delivery date estimates based on a list of holding libraries, which also serves as essential data for the library's collection-building profile and policies. GIST successfully melds two workflows into one and turns a multi-step process into an either/or decision executed from a single screen of information.

When the request is made, some of the data from the request interface may automatically trigger decisions to direct requests to ILL or acquisition processes and special statuses. This level of automation helps streamline work and reduce unnecessary delays. The staff's request processing and decision making is empowered by a wide range of possible ILLiad addons (ILLiad web integrations that can be applied in customized acquisitions and interlibrary loan workflow). Mark Sullivan wrote most of the ILLiad addons, which are simple and powerful ways to automate getting web services to work within the ILLiad client and workflow. For example, when staff open the request for details, an ILLiad addon can automatically search for additional information about the item in various web services: Amazon.com, YBP Gobi, OCLC Connexion, etc. The GIST Acquisitions Manager was developed as the most sophisticated addon yet, because it manages collection building funds, records expenditures, and adds automated recommendations based on your Conspectus and how widely the title is held. It accomplishes this using an addon with two sections working together: the Acquisitions Addon is designed to be the item, budget, and Conspectus view, while the Purchase Addon shows where to purchase the title from your default vendor, or from among 20 other vendors.

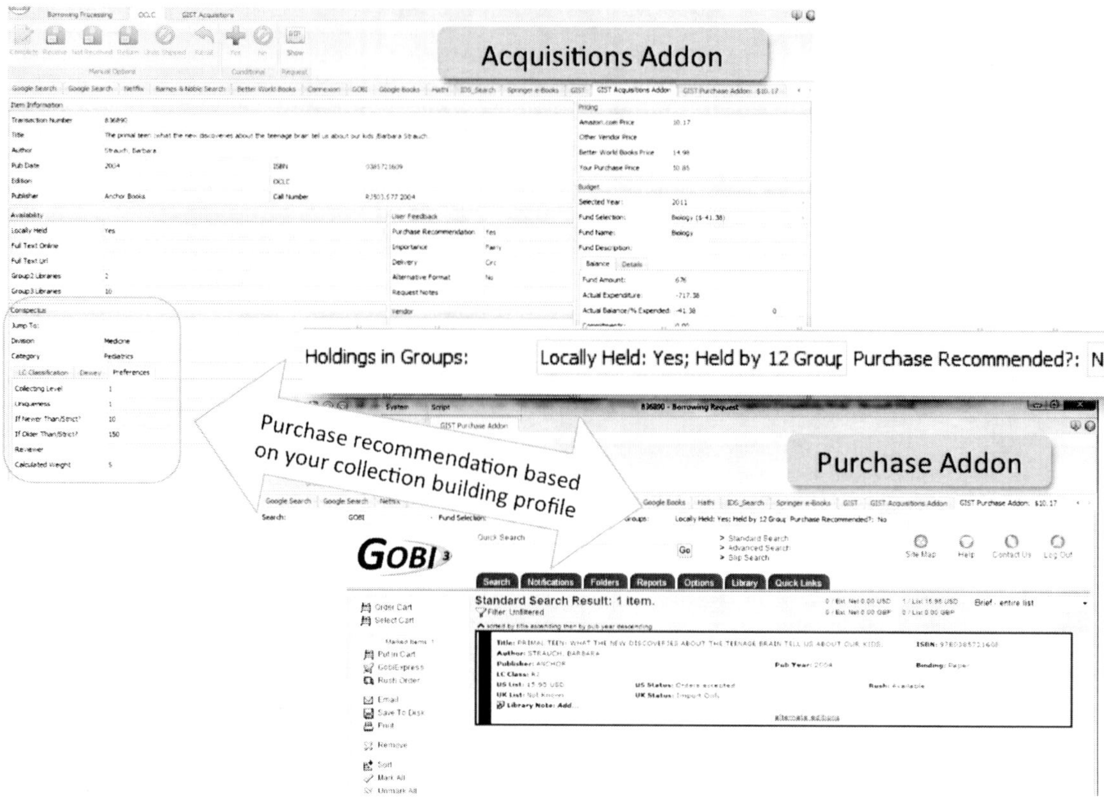

Figure 2. GIST for ILLiad can make recommendations based upon a customized collection building profile.

To leverage useful data in decision making, each user must be able to see key information for each decision in their workflow; the reader evaluates the book and the best way to get their book request fulfilled, and the library staff determine the best value to fulfill a request.

Moving from ILL and acquisitions to gifting and weeding: Automating collections, evaluation, and processing

During the initial GIST project, we learned quite a bit about the value of accessible data sets integrated into a streamlined workflow. Late in 2009, after an experience that showed how labor intensive the evaluation and weeding of some reference collections is, and realizing the similarity of staff processing of thousands and thousands of gift books annually, the GIST team saw another opportunity for automation using what they learned from GIST for ILLiad. By August 2010, we built and released the GIST Gift and Deselection Manager (GDM)—a standalone software that streamlines and automates the functions of gift processing and collection evaluation. Here were two siloed workflows with much in common that were taking precious time away from libraries, and no vendor had developed software with a workflow that made sense of this work. Mark Sullivan's talent for programming and problem solving was put to the test. In more than 20,000 lines of code, the standalone software successfully re-conceptualized how the GIST Team streamlined collection evaluations.

Whether a book is being evaluated for acquisitions or de-accession, GDM automates information gathering and analysis. The process is simple: Wand in the book ISBN and instantly the LC call number or Dewey number identifies the category of the book and whether it fits one of the categories or groups of categories that the library has defined as being desirable. GDM also shows its availability in area libraries, resale market value, full-text holdings, retrieves local and group OCLC holdings, and makes a recommendation—all in three seconds.

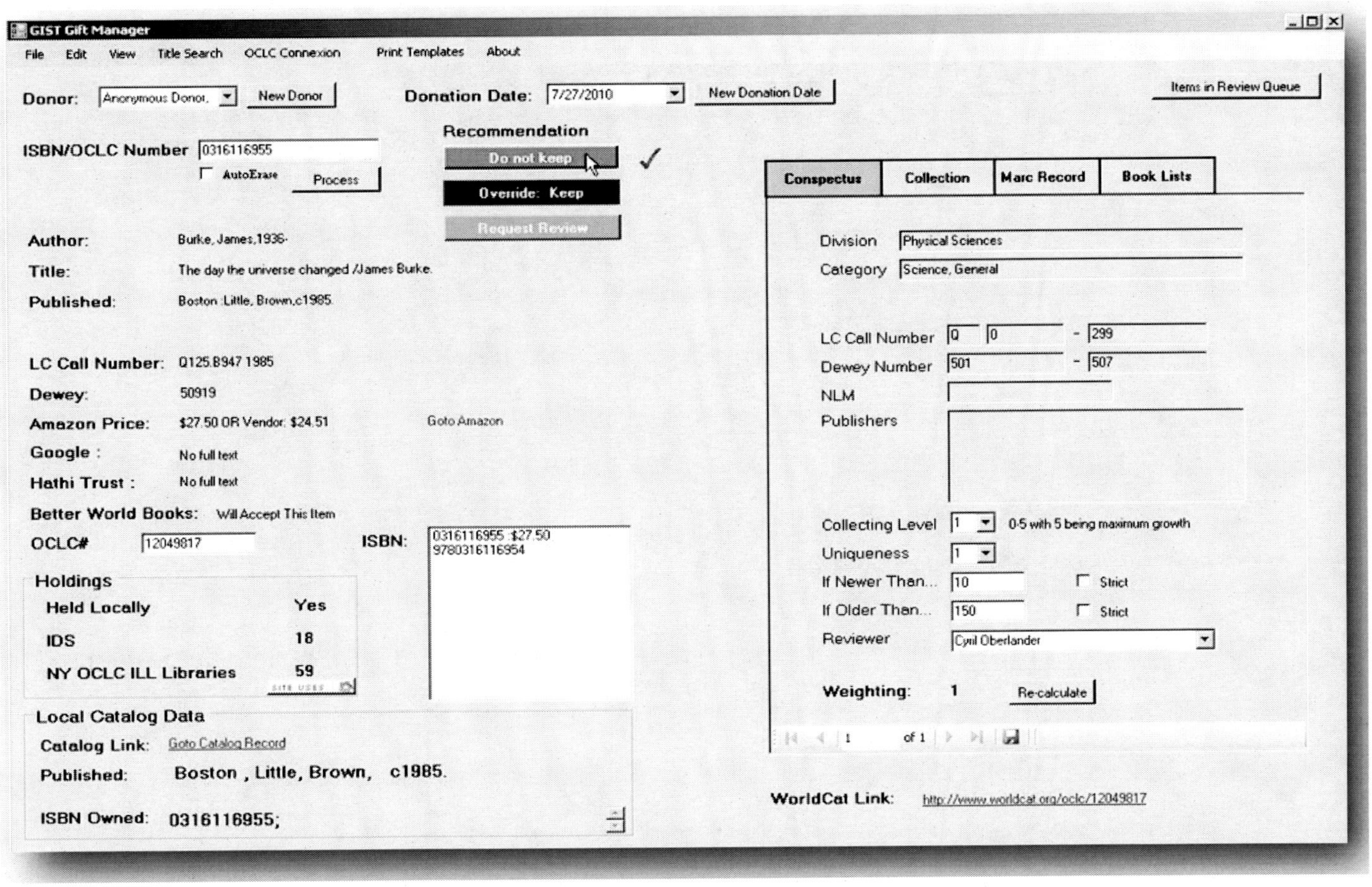

Figure 3. The GIST Gift & Deselection Manager.

At the Milne Library, our staff consumed enormous amounts of time to look over and evaluate thousands of gift books by hand. In our evaluation of more than 4,300 gifts received over 13 months using GDM, we automated the evaluation process and accepted 716 gift books matching our customizable collection-building profile all of which had a net value of $49,000.

The GDM system enables staff to quickly process gift books by determining how rarely the title is held by resource sharing partners, how the title fits in your collection building profile, and whether or not to add the title to the collection—all in a matter of seconds. If keeping the item, staff can immediately move directly into OCLC cataloging with no search needed. If processing a large donation, simply click "print a thank you letter" to acknowledge the donor and the work is done with no need for manually filling in spreadsheets or typing letters.

GDM does more than evaluate gifts. It also evaluates items for de-accession, either individually or in batch. Increasingly, libraries are weeding to make more space for study areas and offices. They either do one search at a time, per book, or they simply make a decision based on items with zero checkouts for 5-10 years. With GDM, the staff can import thousands of ISBNs or OCLC numbers from their zero checkout reports, and GDM will provide a spreadsheet with the key evaluation information in columns. From there, it is simple to sort or filter the data to make the weeding decisions. This is a very powerful system for libraries, as more and more are being asked to weed, while also being asked to add new services. Never before have they had anything that makes the decisions and process so easy. GDM's automation has also proved very useful for identifying rare books to digitize and offer as print-on-demand reprints, as well as for discovering textbooks in your collection to place on reserve. Automated batch processing of a collection is essential to save staff time for more important work that is emerging in new library roles and services.

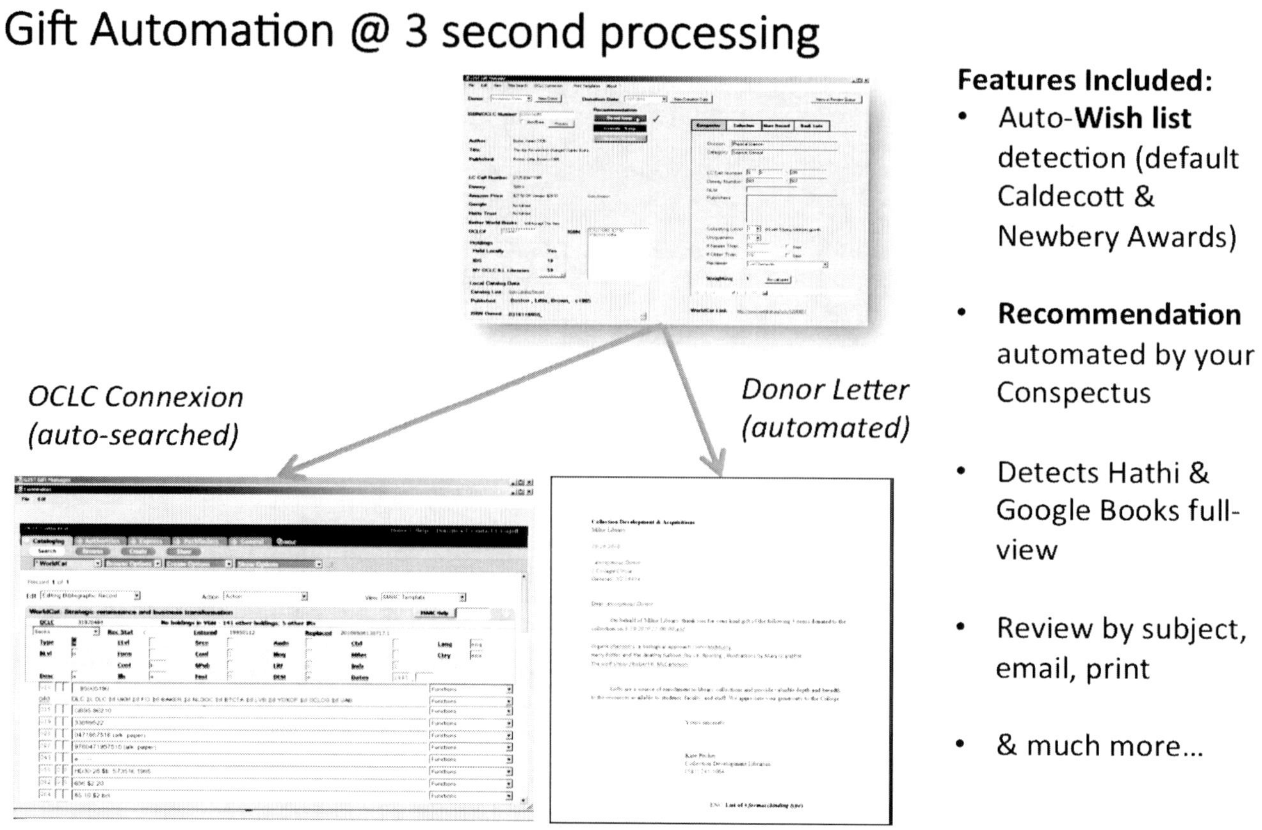

Figure 4. The GDM automates much of the process for managing gifts.

GIST for ILLiad and GIST GDM both incorporate the OCLC Conspectus into a decision making model because it offers a way to divide the subject of requests into divisions that match the scope of institutional collection profiles. The Conspectus was developed by the Research Libraries Group in the 1980s as a way to inventory and categorize research library collections and is being used more frequently to identify strengths, weaknesses, and perhaps even cooperative collection strategies. For a brief history, see: http://www.oclc.org/research/activities/conspectus.html.

By adding the OCLC Conspectus into GIST, we enabled libraries to set their collection-building preferences based on subject. This was very important to the design of GIST, as libraries gave us feedback that librarian review was critical to determining how items fit the library's collection coherence or profile. Because specific title selection was automated at many libraries by approval plans, and new demand-driven e-book packages enable collection profiles, we decided to incorporate the OCLC Conspectus into GIST as a way to empower intelligent automation—having librarians set the parameters of collection-building. For example, using the Conspectus, a library could configure how many agricultural science books were accepted as gifts or purchased by patron driven acquisitions, because the recommendation engine for GIST is looking at the value your library gives that collection building profile.

The OCLC Conspectus is an outline of divisions, categories, and subjects that had been organized by committees; it starts with a subject scheme with divisions and categories, followed by classifications of Library of Congress call numbers, and Dewey Decimal Classification. We simply added "collecting level" and other configurable factors to empower libraries to vary the value placed on titles by subject areas (see table 2).

Table 2. An excerpt of the Conspectus.

DIVISION	CATEGORY	LC CALL NUMBER	DEWEY CALL NUMBER	COLLECTING LEVEL
Agriculture	Agriculture, General	S0-790	630	0
			631.58-631.9	0
			631-631.51	0
Agriculture	Conservation of Natural Resources, Land Conservation	S900-972	333.72	0
Agriculture	Plant Culture	SB0-9999	363.68	1
			631.52-631.57	1
			632-634.8	1
			635	1
			712	1
			714-717	1
			719	1
Agriculture	Forestry	SD0-9999	333.75	0
			634.9	0
Agriculture	Animal Culture	SF0-294.09	636.09	0
		SF311-315	636.1-638	0
		SF360-597	636-636.088	0
			639.7	0
			798.8-798.9	0
Agriculture	Veterinary Medicine	SF600-1100	636.089	2
Agriculture	Aquaculture & Fisheries	SH0-400	639.2-639.6	1
			639.8	0
Agriculture	Wildlife Management	SK351-593	639.9	1
Anthropology	Anthropology, General	GN0-298	599.9-599.97	3
Anthropology	Ethnology. Social and Cultural Anthropology	GN301-699	155.8	3
			301.7	3

The GIST Conspectus design is simple to use and explain. Imagine that every book has an LC or Dewey range, and that it falls into a subject for which you can set a collecting level: 0 = not collecting to 5 = intensively collecting. The collecting level is one important factor among many that allow GIST to recommend buying or keeping a book. One of the other important factors is how widely the item is held among libraries in your consortia, region, or state. The algorithm is outlined in fig. 5:

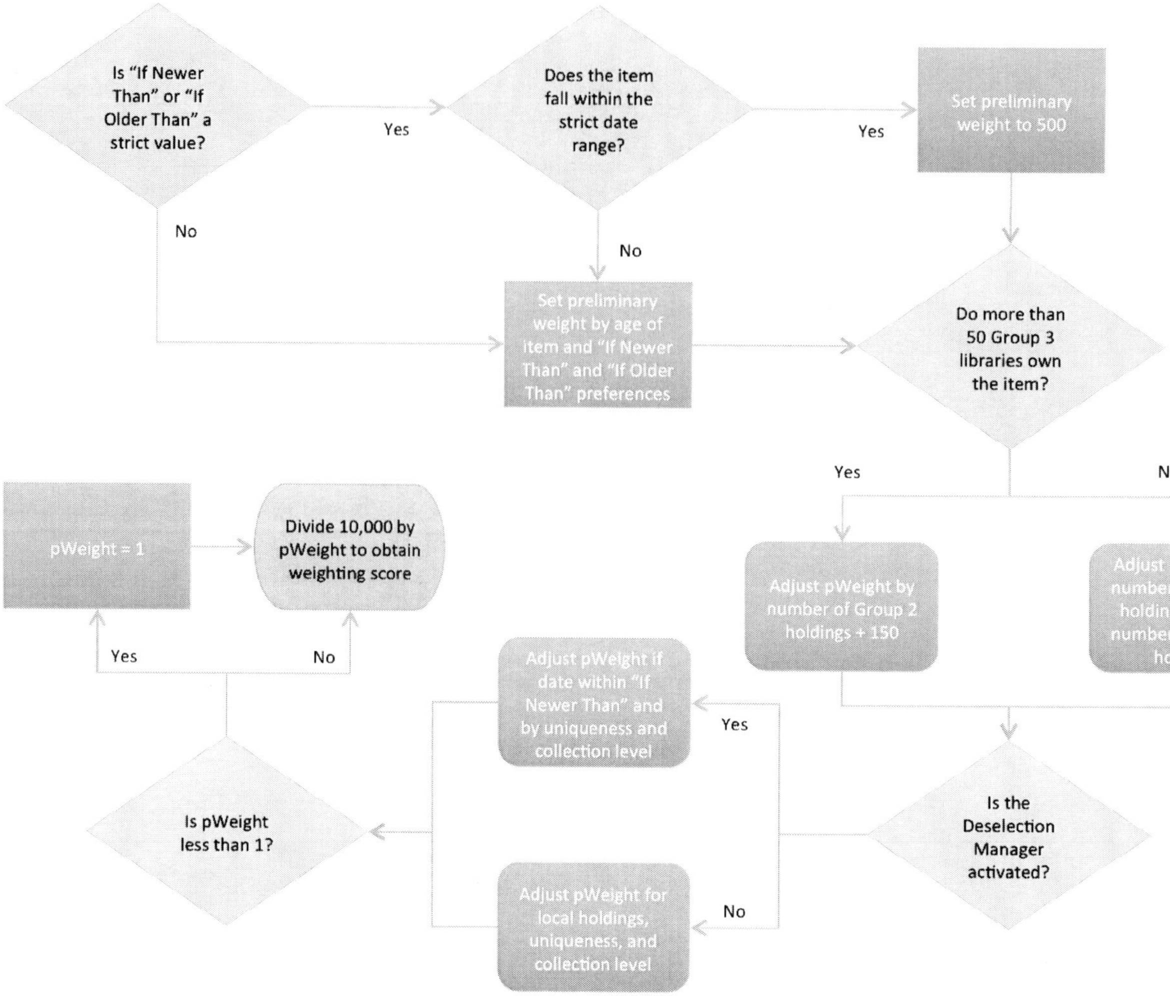

Figure 5. How the GDM assigns weighted recommendations.

Purchase recommendations are determined by the pWeight, which is calculated by several criteria. By default, an item with a pWeight of 100 or more is something recommended for purchasing or keeping. This incredibly important value can be adjusted in the configuration, and is worth adjusting based on the number of libraries in your Group 2 (one of the two customizable groups of OCLC symbols). GIST will likely recommend keeping or purchasing an item if the number of libraries is small, so you can increase the pWeight to adjust accordingly. Another important factor is the Publication Date, as GIST can be configured to recommend items newer or older than a specified number of years. Key to the effectiveness of the Conspectus and this algorithm in GIST is that you can adjust your collection building profile as you process items, making it easy to fine tune your profile as you use GIST.

GIST's ultimate benefit: freeing time for transformation

Today's libraries need new and better tools. More importantly, libraries also need strategies that are going to work for the future and that support the needs of higher education. The overriding question is how can librarians collaborate with faculty in projects that actually matter to the future of higher education? Scholarly and publishing services can provide enormous benefit and opportunities to their institution, supporting digital scholarship or digital projects that enhance collaborative research and development; providing new content distribution models to open access or affordable textbooks; managing data and research curation; and providing instructional design and project management services.

At Milne Library, we are developing our skills around a much needed combination: the ability to help facilitate and curate research and data, in conjunction with delivery via an academic-friendly, content-distribution model.

Digital scholarship

Libraries can directly support the educational mission by providing technical support for digital scholarship initiatives. At Milne Library, we are working with Paul Schacht, Chair of the English department, on Digital Thoreau (http://www.digitalthoreau.org). As the site explains, "Digital Thoreau invites scholars, teachers, students, and all who care about the thoughts and life of Henry David Thoreau (1817-1862) to engage in new and meaningful ways with Thoreau's writings, Thoreau scholarship, and each other."

Digital Thoreau is a digital text of *Walden* that uses Text Encoding Initiative markup (TEI) to build an immersive online environment rich in scholarly annotations, referrals to works held by the Thoreau Society and the Thoreau Institute, and social tools that allow users to create conversations around *Walden*. To make this happen, our wonderful Technical Services staff and librarians learned TEI and have applied metadata to the text of seven editions of *Walden* encompassing some 400 files, plus the works of Thoreau scholars Walter Harding, Ron Clapper, and others. Thanks to Joe Easterly, our Electronic Resources and Digital Scholarship Librarian, as well as Milne's Technical Services, for all of the TEI metadata work to make *Walden* interactive through a Versioning Machine.

New distribution models

One of the most exciting things about the future of library work is collaborating with researchers in shaping new content-distribution and collaboration models. Libraries are now in a position to help students, faculty, authors, and researchers by offering publishing services that range from hosting open-access journals and assisting the publishing of student or faculty work, to hosting repositories and maintaining collaborative e-portfolios of student and faculty research. One of the most promising areas for publishing is open educational resources and, in particular, open textbooks, which are free online textbooks.

Research curation

Increasingly, research is no longer done by a small group of individuals. It is a large collaboration. Who is in charge of preserving this body of work and the data it contains? Who makes it accessible to others? Who can design a collaborative framework for these researchers? Right now, no one does. Librarians are wonderful connectors—they know how to connect people, ideas, and resources. If they are encouraged to develop the skills of project/portfolio management, they can actually develop research teams as a facilitator or project manager—provided they have time.

Milne library staff are finding ways to assist faculty and students in collaborations—from Digital Thoreau to creating Omeka sites for geoscience rock collections and Geneseo's art collection. These efforts extend the works beyond the classrooms or labs, and engage the larger community.

Works cited

Education Advisory Board. 2011. "Redefining the academic library: Managing the migration to Digital Information Services." Washington DC: The Advisory Board Company.

GIST and Patron-Driven Acquisitions

Chapter 1:

GIST for ILLiad

By Tim Bowersox, Access Services Librarian, SUNY Geneseo

Introduction: The case for GIST

The Getting It System Toolkit (GIST) evolved from the realization that the traditional model of just-in-case acquisitions was no longer sustainable. Acquisitions budgets are being pinched by rising serials expenditures and a steady rise in the cost of monographs, which is reflected in an analysis of the Association of Research Libraries (ARL). Between 1986 and 2008, median serial expenditures in ARL libraries rose 374% (Kyrillidou 2009). With the unit cost of monographs also rising 89% during that period, ARL libraries showed no increase in annual monograph purchases (Kyrillidou 2009). As a result, interlibrary borrowing by ARL libraries increased by 295%—by 2008, the ratio of monographs purchased to borrowed was 1:1, compared to an almost 5:1 ratio in 1986 (Kyrillidou 2009).

Putting these pieces together, it is clear that ARL libraries have had to increasingly rely upon other institutions to meet the demands of their patrons. This is corroborated by a 16% decrease in total circulation (Kyrillidou 2009). Since there was no increase in the number of monographs purchased annually by ARL libraries between 1986 and 2008, clearly those that were purchased did not meet the needs of their patrons. This indicates that ARL libraries did not change their collection building strategies to make more efficient use of their shrinking monograph budgets.

The old "80/20 Rule," which states that 20% of a library's collection gets used 80% of the time (Nisonger 2008), has always reflected poorly upon the collection development practices of academic libraries. But as disparaging as the rule is, it may actually be getting worse. A 2011 study by OCLC and the Ohio Library and Information Network (OhioLINK) found that only 6% of the OhioLINK collection circulates 80% of the time (Gammon 2011). Clearly, the traditional selection practice of purchasing monographs just in case patrons need them is unsustainable. Academic libraries can no longer rely on approval plans and painstaking selection to build prescient collections.

The alternative to the traditional model is patron-driven acquisitions (PDA), which places the role of selection squarely upon the patron. GIST employs PDA using a just-in-time workflow, whereby items are evaluated for purchase at the point of need. By tapping into various data sources in real time, GIST is able to automate much of the initial evaluation based upon the Amazon.com vendor price, local and regional holdings, and open access availability. Built as an enhancement to the ILLiad interlibrary loan software, developed by Atlas Systems, Inc., GIST introduces request management to the acquisitions process and provides for a streamlined, collaborative workflow with interlibrary loan. The result is a powerful system that allows libraries to build smarter collections, buying the materials that make sense and borrowing the ones that do not.

The development of GIST

GIST was developed in 2009 at the State University of New York at Geneseo's (SUNY Geneseo) Milne Library, which itself had faced years of declining monograph purchases and an increased reliance upon interlibrary borrowing. Implementing a patron-driven collection development strategy, therefore, became a necessity. After exploring different PDA models, however, GIST's developers realized that more could be done to make the process more efficient.

Data-driven PDA

Data-driven PDA employs retrospective use data from interlibrary loan requests and circulation transactions to identify collection gaps. Also referred to as access-informed collection development, this model is based upon the premise that the attempts of patrons to access information are key indicators of need (Mortimore 2005). By reviewing interlibrary loan data, trends appear that indicate which titles or subject areas are lacking in a collection. This can be reinforced by evaluating trends in circulation data. When compared against holdings data, libraries can target areas for growth and deselection. In this model, titles frequently requested through interlibrary loan become good candidates for acquisition.

This model demonstrates the value of usage data in the collection development process. However, it assumes that past data provide an accurate representation of present needs. Because trends in research and scholarship can change semester by semester, titles and subject areas in demand two years ago may no longer be the focus of students and faculty. All it takes is a focused seminar course or a determined doctoral student to skew the usage data. The data-driven model, therefore, still employs a speculative strategy for collection development.

This speculative strategy is also more expensive than a PDA model that employs just-in-time acquisitions. Because each interlibrary loan request incurs at least some cost, a library is actually paying more than the purchase price of frequently requested titles. At best, even if a library does not have to pay another library for borrowing an item, return shipping must still be paid. At worst, a library could end up paying more to borrow a book once than it would have cost to buy it.

Just-in-time PDA

The just-in-time PDA model solves the problems inherent in the retrospective data-driven collection building model. Just-in-time PDA is also referred to as on-demand collection development, and sometimes misunderstood as Demand Driven Acquisitions, which usually refers to a set of ebook titles loaded into the local catalog for short-term loan or acquisitions based on number of uses. The just-in-time PDA model applies predefined purchase criteria to each interlibrary loan request to determine whether to buy or borrow the item (Ward 2003). These criteria are collaboratively identified by interlibrary loan staff, acquisitions staff, and collection developers, and include such factors as price, publication date, format, language, and subject.

Purchases are typically made from an allocated sum of money, meaning this model is often a supplement to the traditional just-in-case model at many institutions. Although this greatly mitigates the impact of the just-in-time model, it still yielded positive results in the cases of Purdue University and the University of Wisconsin-Madison. In each case, items purchased through the just-in-time model circulated two to three times more often than traditionally acquired books (Ward 2003).

At Purdue and Wisconsin-Madison, staff had to manually research each requested item to determine whether or not it met their purchase criteria. This would involve, among other things, searching multiple data sources. If an item was selected for purchase, that patron's request was then fulfilled outside of the interlibrary loan system. Furthermore, neither Purdue nor Wisconsin-Madison included the patron's recommendation as one of the criteria (Ward 2003), overlooking a valuable source of data for making informed purchasing decisions.

GIST was developed to fill these gaps. By importing price, holdings, and other data into the patron's request, staff had the data they needed to make decisions more efficiently. GIST also keeps the entire request lifecycle within ILLiad, allowing staff and patrons to manage and track a request in a single system. Most importantly, GIST enables patrons to recommend an item for purchase at the point of request, allowing patrons to truly drive the acquisitions process.

The GIST workflow

Because GIST was developed as an enhancement to ILLiad, the overall workflow resembles that of an interlibrary loan request (see fig. 1). Since ILLiad is a request management tool, the workflow begins when a patron submits a request. Based upon a library's chosen criteria, ILLiad can be configured to direct that request to either an interlibrary loan queue or an acquisitions queue. In the latter case, acquisitions staff can review the request, purchase the item, attach OCLC holdings, encumber funds, and notify the patron all from within ILLiad. If the item was not purchased, acquisitions staff can simply route the request to the interlibrary loan queue to be borrowed—an entirely seamless experience for the patron.

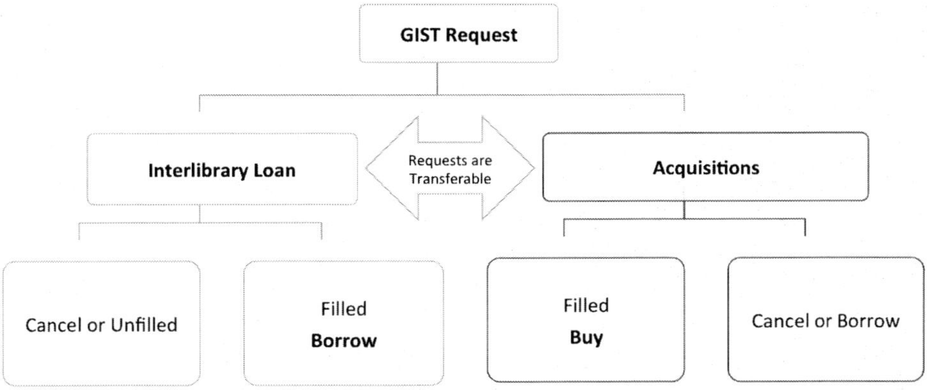

Figure 1. The GIST workflow, showing the seamless connection between ILL and Acquisitions.

The ILLiad web pages

Although staff members have the ability to manually input patron requests, ILLiad provides patrons with an online platform for placing and tracking their requests. Libraries can customize this platform, commonly called the ILLiad web pages, both in terms of style and functionality (see fig. 2). GIST takes full advantage of this capability to provide an enhanced patron interface, while also importing data behind the scenes to facilitate informed decision making.

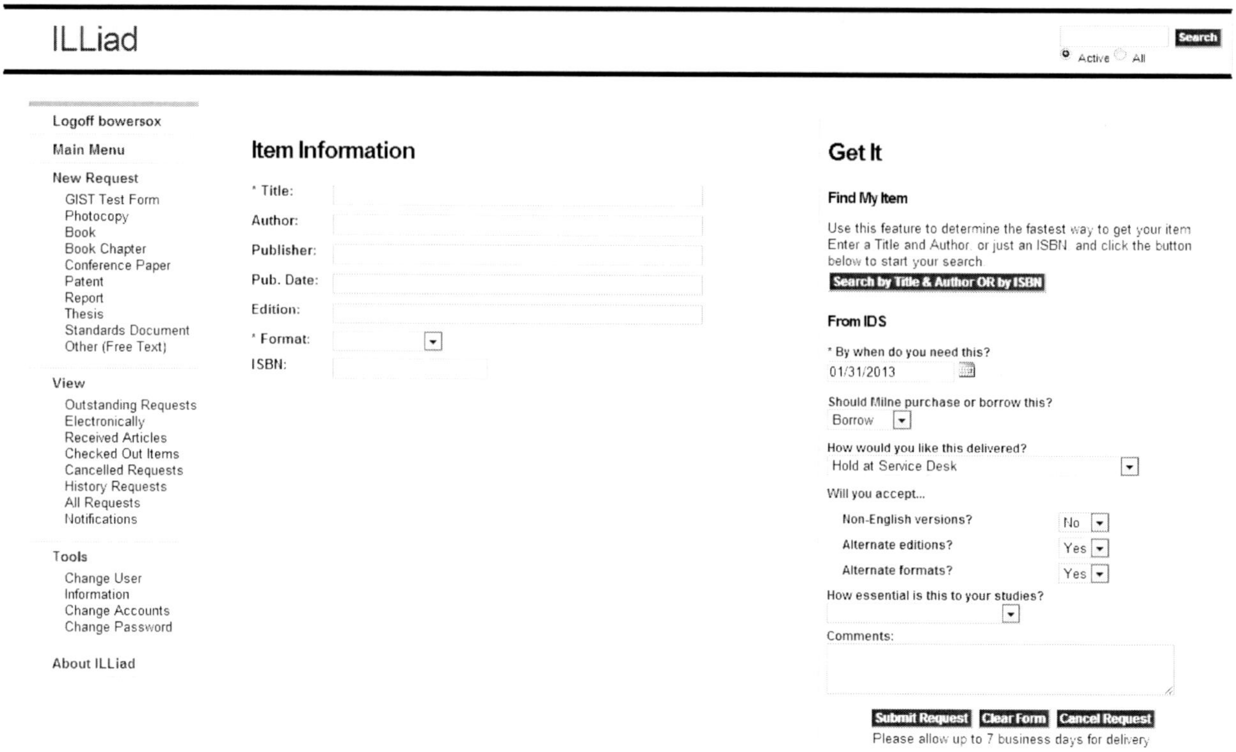

Figure 2. A blank ILLiad request form using GIST.

Amazon.com reviews

By leveraging data from Amazon Web Services, GIST provides patrons with user and editorial reviews from Amazon.com (see fig. 3). This information can help patrons determine whether or not a requested item may prove useful. By linking directly to the item at Amazon.com, patrons can also take advantage of Amazon's "Look Inside" functionality (when available). Both of these resources can help patrons evaluate whether or not an item is worth requesting.

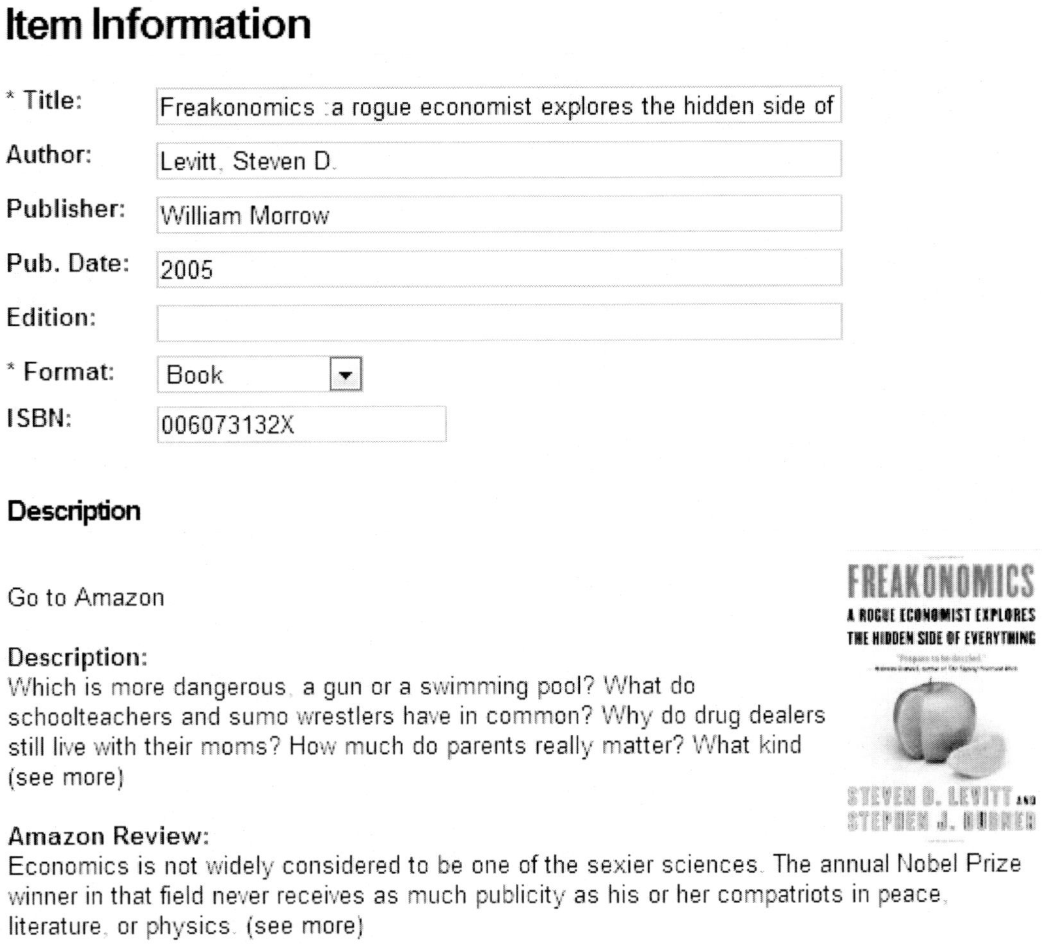

Figure 3. Patrons can see item descriptions and reviews from Amazon.com.

Open access availability

In an effort to improve the discoverability of open access books, GIST provides users with the availability of their requested item in Google Books, the HathiTrust, Project Gutenberg, and the Internet Archive. If an item is not available in full text, often a patron can preview the item and its table of contents in Google Books (see fig. 4). All of this is presented prominently on the page, encouraging patrons to first explore this method of access before requesting a physical copy. Not only does this provide instant access to certain titles, it also reduces the workload for staff when it comes to public domain and open access works.

Figure 4. Patrons can see viewing options from Google Books and open access sources, if available.

Local and regional library holdings

GIST uses OCLC's WorldCat Search API to leverage library holdings data. If an item is held by the patron's library, then a notification appears on the page along with a link to the item in their library's catalog (see fig. 5). This improves the discoverability of library materials, while also encouraging patrons to make good use of their library's collection.

Figure 5. GIST connects patrons to locally held items.

If a requested item is not locally held by the patron's library, GIST uses regional holdings data to provide an interlibrary loan delivery estimate (see fig. 6). Libraries can configure GIST with two custom groups of OCLC symbols—ideally consortia or statewide libraries—and the number of days it typically takes to receive materials from each group (for example, three days for consortia and seven days for statewide libraries). When GIST determines that libraries in a group hold the requested item, the patron is presented with the corresponding delivery estimate. In this way, the interlibrary loan process becomes more transparent by providing information patrons have come to expect from online retailers.

Figure 6. GIST can provide custom delivery estimates for ILL.

Behind the scenes, GIST stores the holdings data in the submitted request. As explained in more depth later in this chapter, this data can be used to automatically determine whether or not a requested item is eligible for purchase. The data are also used to automatically route requests into specific workflow queues. Obviously, many libraries do not want to purchase multiple copies of an item they already own. But some libraries may not want to purchase an item if a certain number of copies are held in their consortia or state. In such cases, it may make the most sense to borrow an item in order to facilitate the development of a more diverse regional collection.

Purchase prices

In the spirit of providing patrons with all available options for accessing their requested items, GIST lists the lowest Amazon.com and Better World Books purchase prices (see fig. 7). Although likely the exception to the rule, patrons may be prompted to order their own copy of a requested item if the price is reasonable. These days, it is not uncommon to find some used paperbacks for under $5, so it could be beneficial for patrons to buy a copy, especially if they expect to need it for a long period of time.

Figure 7. GIST informs decision making by providing purchase prices.

What is more likely, however, is that faculty or subject selectors would use the price in evaluating whether or not an item should be purchased by the library. By weighing the price against their current acquisitions budget, selectors could opt to borrow more expensive items to save money. This is especially useful when also considering the number of group holdings. In other words, if a book costs $100 and 10 consortia libraries own it, then it is likely not worth purchasing a local copy.

GIST records the purchase price at the time of request, since this is one of the most important criteria that can be used in PDA. As explained later in this Chapter, this data can be used to automatically determine whether or not a requested item is eligible for purchase. Because many libraries place a cap on the purchase price of an item (whether it be $50, $100, or $150), this can ensure that only eligible requests reach the acquisitions staff. Different criteria can even be used based upon the patron's status (faculty, undergraduate, etc.) and/or department, such that faculty requests could have a higher cap than students (or science faculty members have a higher cap than those of humanities faculty).

Patron recommendations

Providing patrons with as much useful data as possible enables them to become informed requestors. Patrons can use the information provided by GIST to not only determine whether or not to request an item, but also to make a sound purchase recommendation to their library. After all, there is no better indicator of patrons' needs than the patrons themselves. Because ILLiad allows patrons to record comments in each request, patrons can provide context to their recommendations.

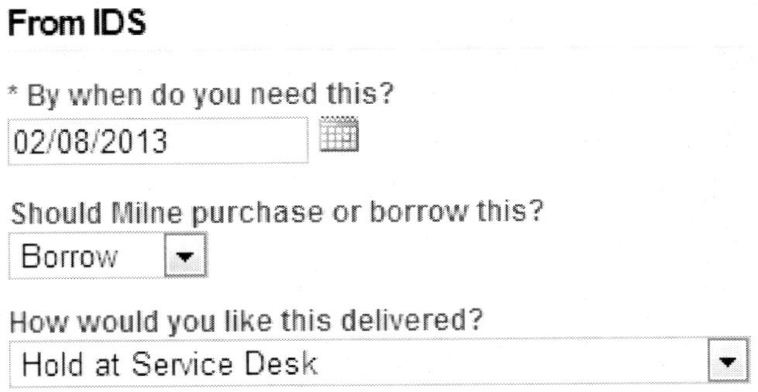

Figure 8. Patrons can recommend an item for purchase at the point of request.

With GIST, patron recommendations become one of the criteria used to evaluate whether or not an item should be purchased. This allows staff to weigh qualitative feedback against price, holdings, and bibliographic data. Because GIST does not purchase items automatically, libraries still maintain full control over the final purchasing decisions. These decisions just become more informed by bringing the patrons into the process.

Status-specific web pages

ILLiad allows libraries to create personalized web pages for different segments of patrons. Called "status-specific web pages," these can be customized based upon a patron's user account status. For example, this means that undergraduates may not have access to the same web pages as faculty members. Similarly, the ILLiad web pages may look entirely different for undergraduates than they do for faculty members.

In the context of GIST, this means that libraries can limit the scope of GIST for different types of patrons. At the most extreme, a library may only allow faculty requests to be considered for purchase. By making a GIST-customized page for patrons with a user status of "Faculty," all other patrons will still see the default ILLiad web pages. Alternatively, libraries may not want the purchase prices to appear for undergraduates. By using a GIST-customized page for patrons with a user status of "Undergraduates," this becomes possible.

Use with or without OpenURL

OpenURL is one of ILLiad's most powerful features. This feature can create patron requests based upon OpenURL strings sent by electronic resources. This spares patrons the time of hand keying the bibliographic information for items they want to request. Upon authentication, patrons simply have to set their preferences and submit the request.

ILLiad can be configured to use a GIST-customized web page for requests received via OpenURL. For books and media found in WorldCat and other electronic resources, this allows patrons to view the data provided by GIST and make purchase recommendations. This is not only easier for patrons, but it can also provide libraries with a more diverse range of titles for purchase consideration.

For requests not submitted via OpenURL, GIST provides users with a WorldCat search widget built right into the ILLiad web page that allows search by title and author. This encourages users to more easily find the items they need, especially if they are unsure of the exact titles or authors. Each time a patron selects an item from the search results, he or she is presented with the relevant Amazon.com reviews, open access availability, holdings, and price data. Not only does this allow patrons to make a more informed decision, it also provides staff with precise bibliographic data—usually an uncertainty with non-OpenURL requests.

ILLiad automation: the database, routing rules, and custom queues

Once a patron submits a request online, it is added to ILLiad's database as a transaction. Each transaction contains the item's bibliographic information as well as any preferences provided by the patron. Each transaction is linked to the patron's user account and is identified by a unique accession number, such that each transaction can be identified by the item requested, who requested it, and the transaction number.

The data imported from the ILLiad web pages are stored in one of several database tables installed with GIST. This table, named GISTWeb, is linked to each transaction by the transaction number, such that each transaction is associated with a purchase price, a local holdings indicator, group holdings counts, an open access availability indicator, and a patron recommendation.

ILLiad can be configured to use this data to determine whether or not a request is treated as an interlibrary loan. By creating SQL queries called "routing rules," ILLiad can assign a transaction to a particular queue based upon a set of criteria. Routing rules can evaluate transaction, GIST, and user data, allowing libraries to be as broad or specific as they would like. Transactions that meet a library's baseline purchase criteria will be routed to a custom queue for new purchase requests. Those that do not will be routed to a default queue for new interlibrary loan requests.

Each queue in ILLiad represents a transaction's status in a particular workflow. GIST employs custom queues dedicated to the acquisitions workflow, each one corresponding to a particular step. New requests are contained in a queue called "Awaiting Purchase Request Processing" that is monitored by acquisitions staff. Once an item has been purchased, it is then routed by staff to a queue called "Purchase Request Ordered from Vendor." When a purchased item arrives, staff need only browse that queue to find the transaction to which it belongs.

Queues are also important for patrons. The name of each queue serves as a transaction's status, and that status can be seen when tracking a request in the ILLiad web pages. This adds a level of transparency to purchase requests that would not be available without ILLiad.

Request management: ILLiad client addons and custom emails

Staff members manage patron requests using the ILLiad client software. Each staff member has his or her own account, allowing for a customized experience. For example, acquisitions staff members do not need access to the interlibrary loan queues. These can therefore be hidden, providing an uncluttered view. Individual queues can also be hidden, allowing for a very focused and easy to learn interface.

Each transaction opens in its own window, presenting staff users with an array of tabs. The "Detail" tab presents the bibliographic and user data for the transaction, as well as the notes left by the user or added by staff (see fig. 9). The "History" tab allows staff to track changes to the request, which can be helpful for determining when a request was originally submitted, who last worked on it, and so on. The "Z39.50" and "OCLC" tabs allow staff to search for the requested item in their local catalog and WorldCat, respectively.

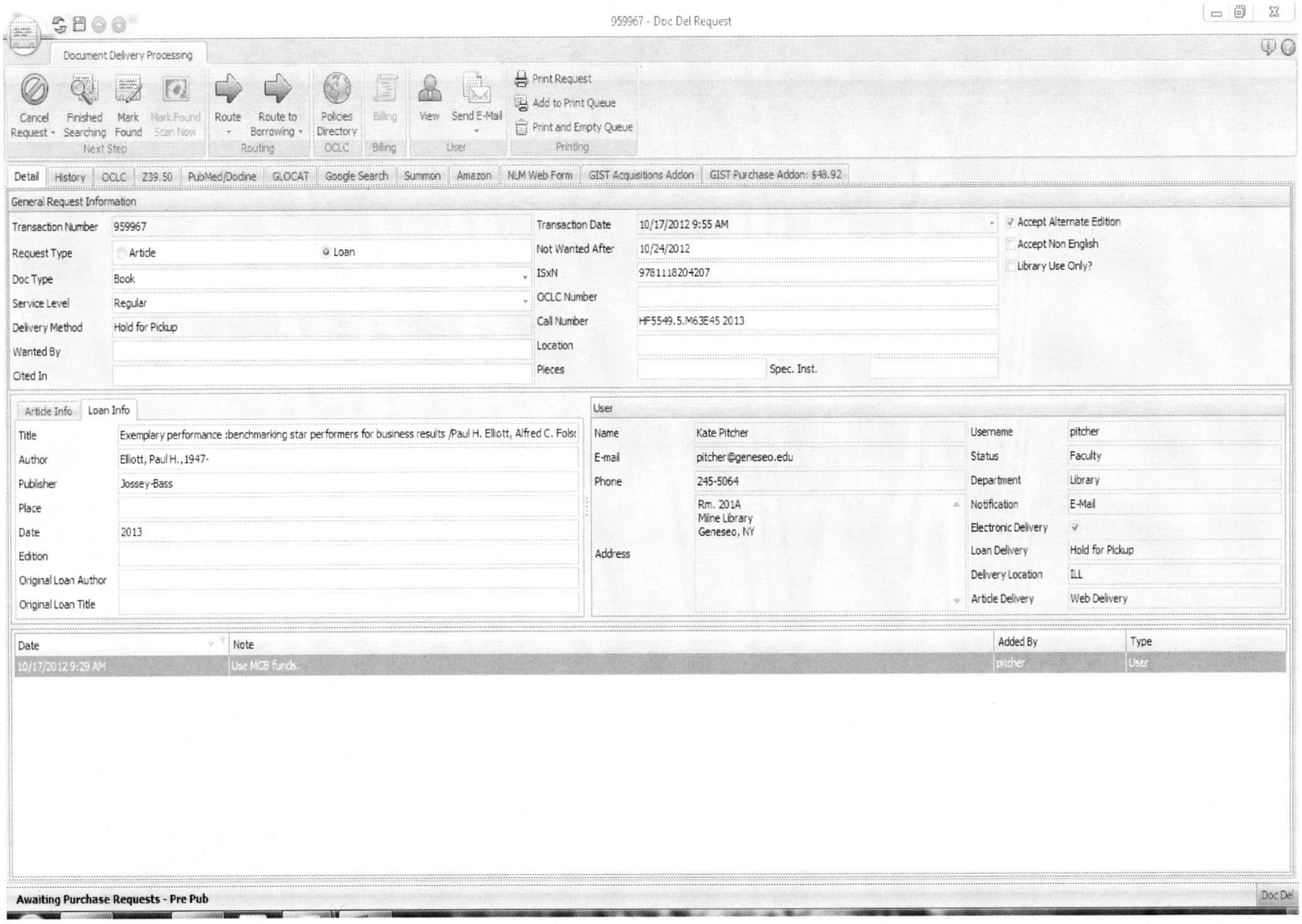

Figure 9. ILLiad presents bibliographic and other request information in the Detail tab.

GIST provides additional tabs through enhancements called "addons." Similar to the way browser extensions add functionality to Google Chrome or Mozilla Firefox, addons are third-party extensions that expand ILLiad's functionality. In technical terms, addons are scripts written in the open-source Lua programming language (http://www.lua.org). This is a simple, lightweight language that allows addons to interact with ILLiad. Atlas Systems provides a directory of available addons that users can easily download and install, while also providing documentation to help developers modify and create addons (https://prometheus.atlas-sys.com/display/ILLiadAddons). Because addons are open source, even non-programmers can create addons by modifying and expanding upon existing ones (with attribution, of course).

Many addons enable ILLiad to interact with websites. For example, the National Library of Medicine or NLM Form addon (developed by California Polytechnic State University) loads the NLM's interlibrary loan request form within an ILLiad tab. It allows staff to import their NLM account information (defined in the addon's settings) along with the requested item's bibliographic information from the ILLiad transaction—all with one click. This means a task which would have taken several minutes, now takes seconds.

The GIST development team has created several addons to create an efficient acquisitions workflow. The "GIST Purchase Addon" allows staff users to shop for requested items from vendors including Amazon.com, Barnes & Noble, Better World Books, and GOBI, among many others (see fig. 10). The addon automatically searches each vendor's website for the requested item, sparing staff from repeating manual searches in multiple browser windows. Because the page is opened within the ILLiad client, the addon can actually capture data from the web page.

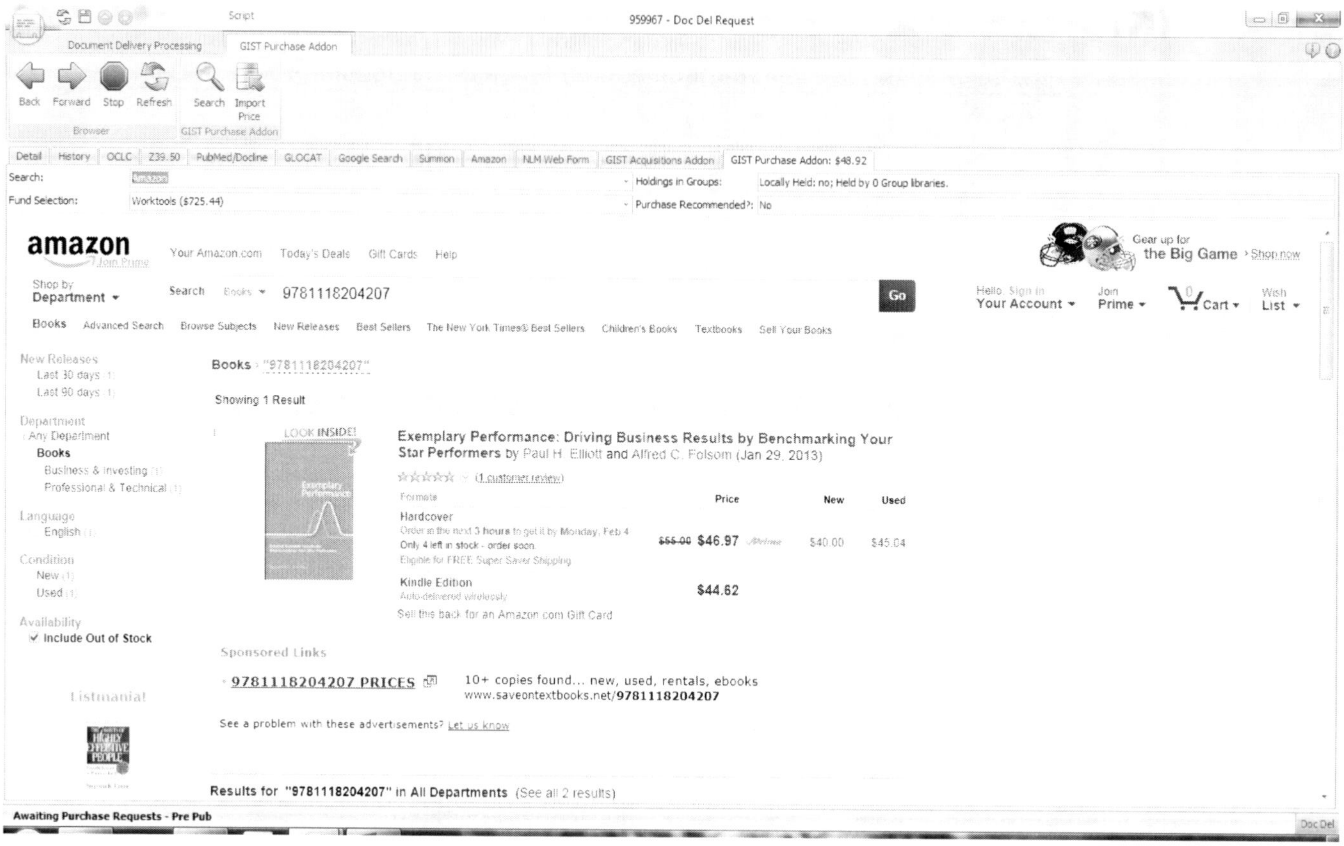

Figure 10. The GIST Purchase addons allows staff to order items from within ILLiad.

This captured data is imported to the "GIST Acquisitions Addon," which serves as an order record for each transaction. Data about the order, including the vendor name, purchase price, shipping cost, and date, are imported into the addon. Once verified, staff can update the order, which then deducts the amount from the fund associated with the patron.

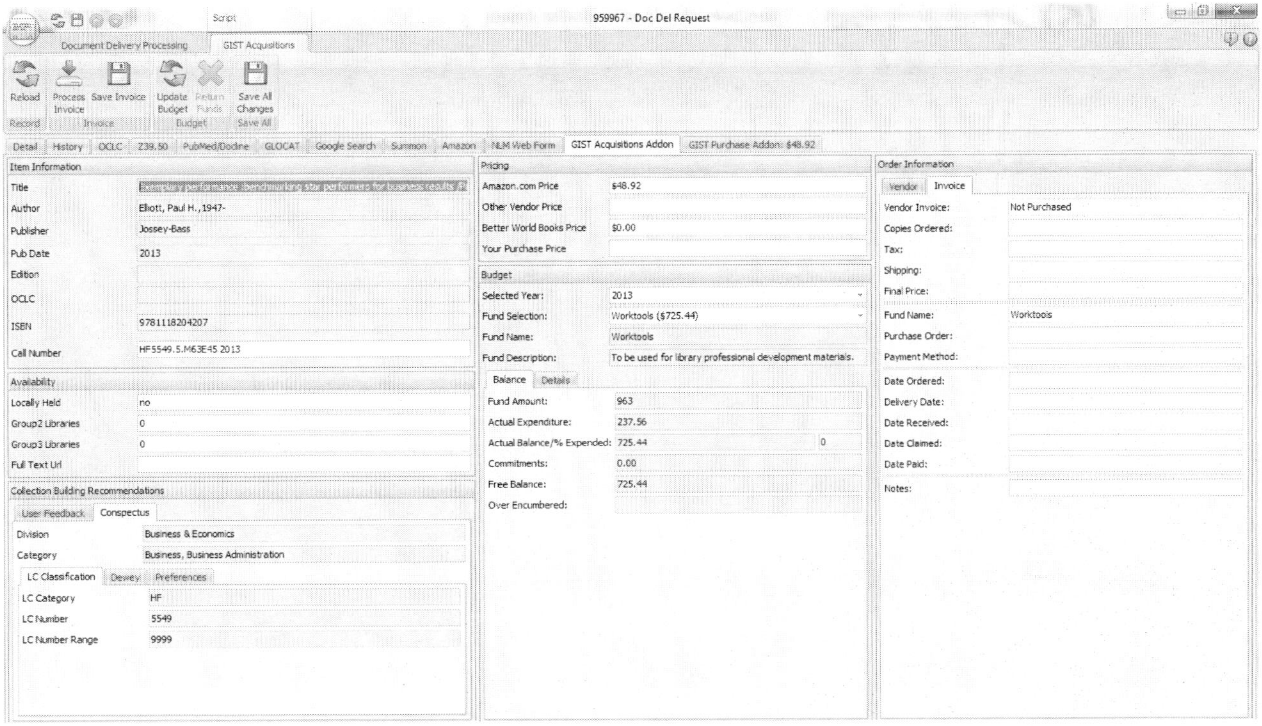

Figure 11. The GIST Acquisitions addon displays information about the order.

Acquisition funds and vendor information are managed in a separate "GIST Database Addon," which appears in the main ILLiad client window. Funds are created by fiscal year and can be associated with specific departments or users. When staff users process a purchase request, the fund associated with the patron is selected by default, although it can be overridden with each request. (For example, a fund called "Biology" could be created and associated with the "Biology" department, such that the "Biology" fund will be selected by default for all purchase requests made by patrons in the Biology Department.)

The GIST Database Addon allows acquisitions staff the ability to track detailed fund balances, including encumbrances and overages (see fig. 12). Allocations can be modified to add or remove money from funds throughout a budget year. In addition, funds can easily be copied from one year to the next.

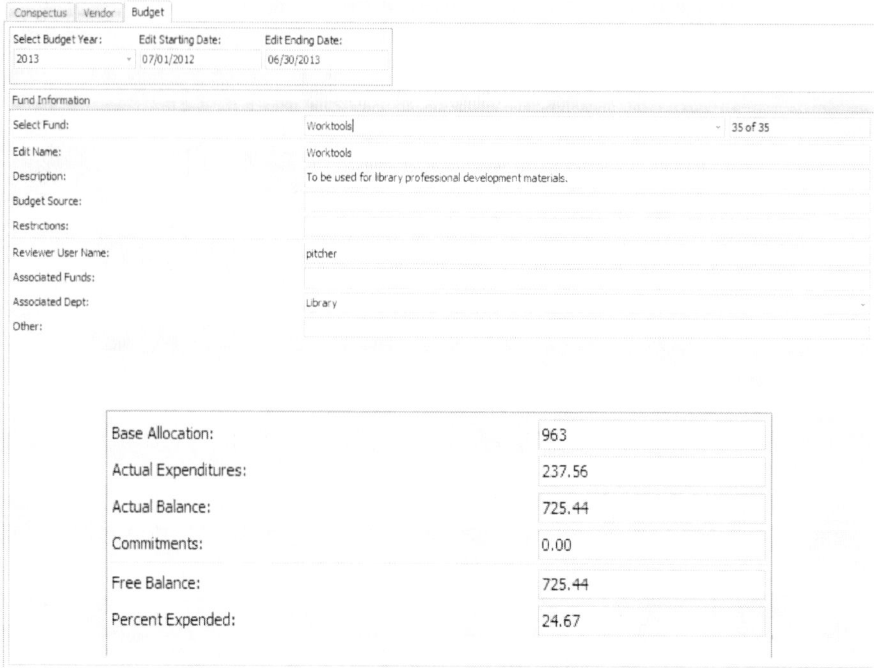

Figure 12. Funds can be managed from the GIST Database addon.

Vendor data can also be managed in the GIST Database Addon (see fig. 13). GIST comes preloaded with data on more than 20 vendors—the same vendors available in the GIST Purchase Addon. This allows staff to hit the ground running once GIST is installed, though they do have the option of adding additional information such as discounts, sales tax rates, contact information, and more. This can make contacting vendors about particular orders easier, as the data are also displayed in each request's GIST Acquisitions Addon.

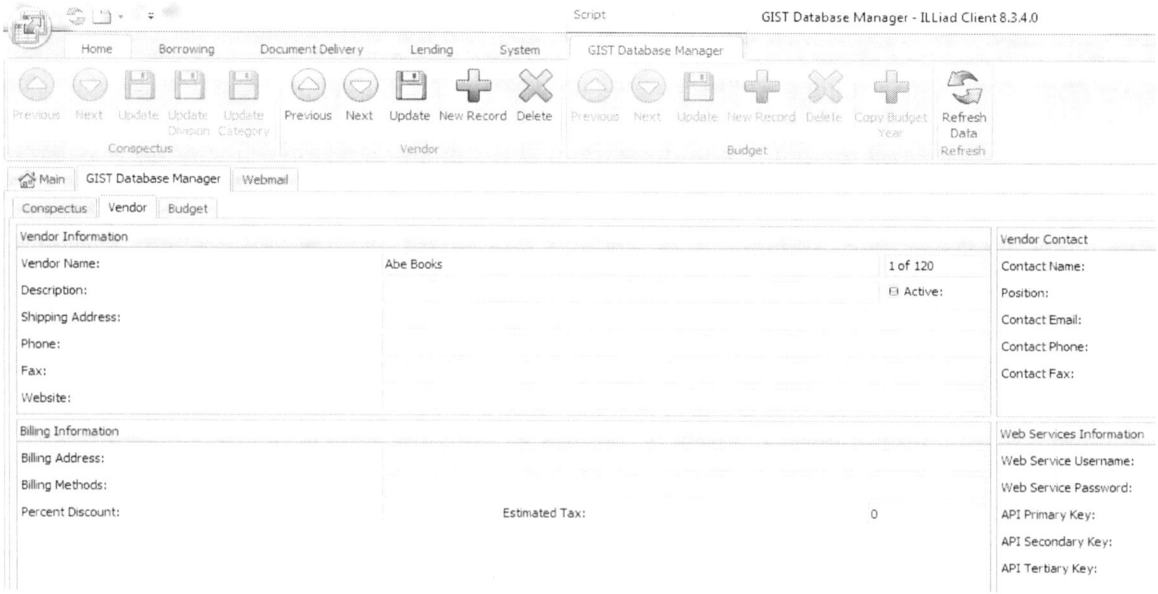

Figure 13. GIST for ILLiad comes preloaded with 20 popular vendors and can be customized within the Database addon.

Also present in the GIST Database Addon is the Conspectus (see fig. 14). This is a customizable collection building profile that can generate weighted recommendations that display in the GIST Purchase Addon. This gives staff insight into whether or not a requested item fits within the scope of their collection. Health sciences libraries, for example, would place more emphasis upon acquiring medical textbooks than a law library. Knowing this, staff users are more inclined to make an informed purchase decision.

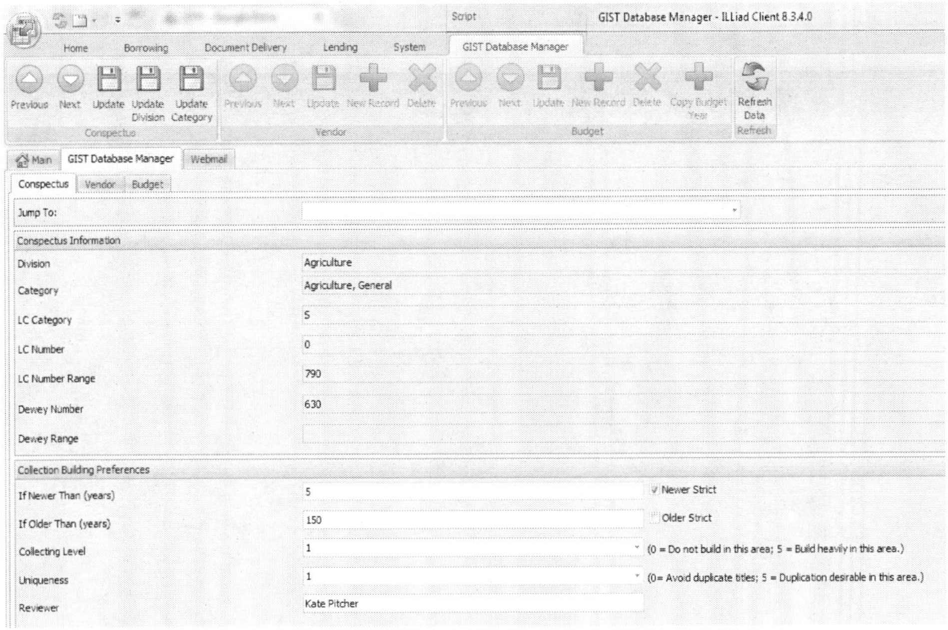

Figure 14. GIST can provide weighted recommendations using the Conspectus, a collection building profile.

Once an item has ultimately been purchased, an optional OCLC Connexion addon allows staff to attach their library's OCLC holdings to the item's record and download it to their library catalog (see fig. 15). After the item has arrived and been processed, staff can send patrons email notifications directly from within the ILLiad request. ILLiad allows libraries to create any number of customized email notifications, each based upon a text template that can automatically insert transaction and user data. In other words, staff users no longer have to hand type or copy and paste emails to let patrons know a purchased item is ready for pickup.

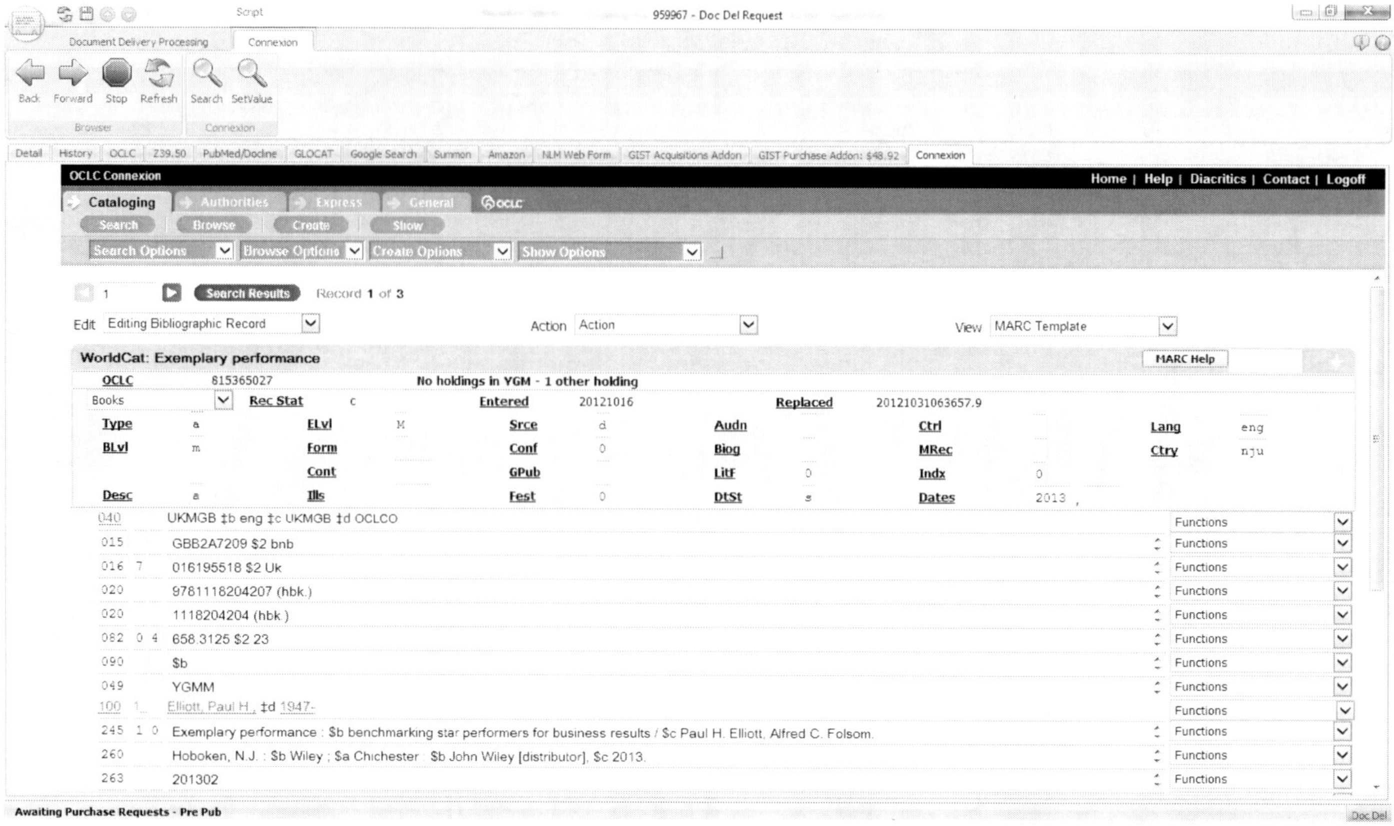

Figure 15. The optional Connexion addon allows staff to add holdings and download records at the point of order.

Impact of GIST on SUNY Geneseo

Prior to GIST, collection development at SUNY Geneseo was primarily faculty driven. Milne Library did not have dedicated bibliographers, though librarians would also place purchase requests. Approval plans were also not in wide use.

Purchase requests were managed in a paper-based system. Though there was an online form in place, it was just one of a few ways faculty and librarians could place requests, including in person, over the phone, via email, on scraps of paper, or even by sending catalog clippings. Even the online form generated emails that were printed out. Regardless, all requests ended up in binders and were attached to order receipts, printouts from the ILS, and any correspondence collected throughout the request lifecycle. Once an item was received and processed, staff would need to hand-enter email notifications to the requestor.

One of the most important impacts that GIST had on SUNY Geneseo was to make the entire purchase request process more efficient. Faculty and librarians need only go to one place for making requests, and purchase requests could actually be tracked online for the first time. Acquisitions staff no longer needed to keep extensive paper records for each request, since much of the necessary data were stored in ILLiad. This allowed the one FTE staff to spend less time keeping records and more time actually fulfilling the requests, improving turnaround time. Thanks to ILLiad's email notification capabilities, routine correspondence was now only a click away.

From 2010 through 2012, acquisitions staff reviewed 8,283 purchase requests (see table 1). These requests consisted of 6,121 items recommended explicitly by patrons. The remaining 2,162 were recommended either by interlibrary loan staff or were considered because they were new releases (which are difficult to borrow through interlibrary loan). Overall, 90% of requested items were purchased and delivered within an average of 10 calendar days.

Table 1. GIST purchase requests at SUNY Geneseo, 2010-2012.

Year	Patron Recommended	Non-Patron Recommended	Total Recommended	Purchased	Turnaround (in Days)
2010	2,961	276	3,237	2,873 (89%)	9.42
2011	1,928	925	2,853	2,612 (92%)	9.86
2012	1,232	961	2,193	1,949 (89%)	10.74
Total	6,121	2,162	8,283	7,434 (90%)	10.01

GIST also provided a means for students to participate in collection development for the very first time. Because students place the overwhelming majority of the interlibrary loan requests at SUNY Geneseo, Milne Library finally had a means for tapping into this valuable source for patron-driven acquisitions. This would in turn allow Milne Library to develop a collection more representative of its patron population. Although the faculty continued to be the primary source of purchases, undergraduates accounted for 29% of all purchase requests (see table 2). If you discount those requests made by librarians, undergraduates actually placed 605 more purchase requests than the teaching faculty.

The impact of GIST upon Geneseo's collection has also been positive. From 2010-2012, Milne Library added 64% more items to its circulating book and media collections than in the previous three-year period (see table 3). When adjusted for the number of years in circulation, items purchased with GIST circulated 25% more often than items purchased in the previous three-year period. Based upon this data, one can see the impact that a wholly patron-driven approach can have upon collection usage.

Table 2. GIST purchase requests by patron department and status, 2010-2012.

Department	Faculty	Graduate	Guest	Staff	Undergrad
Abbey			2		
Administration				63	1
Anthropology	41			22	38
Art	20			3	47
Biology	28	1			162
Business	24			11	234
Chemistry (Biochemistry)	8				32
CIT				7	
Communication & Speech Communication	19			10	58
Communicative Disorders & Sciences, Speech Pathology	16	3			31
Computer Science	40				13
Education	263	20		16	163
English	91	2		1	284
Foreign Languages	41	1			55
Geography	54				23
Geology	7				9

History	318				431
HPE					1
Library	3,245			742	13
Math	146			1	22
Performing Arts, Drama and Music	129				24
Philosophy	67				26
Physics	17			1	23
Political Science (Includes International Relations)	117				170
Psychology	160				165
Sociology	151				42
Undecided	1			1	296
	5,003	**27**	**2**	**878**	**2,363**

Table 3. Circulation of items purchased before and after GIST.

	Pre-GIST			Post-GIST		
Calendar Year	2007	2008	2009	2010	2011	2012
Items Added to Collection	1,723	1,344	1,396	2,383	2,663	2,284
Loans & Renewals as of 12/31/12	5,247	3,448	6,530	5,896	4,703	2,322
Avg. Loans & Renewals per Year	875	690	1,633	1,965	2,352	2,322
Avg. Loans & Renewals per Item	0.51	0.51	1.17	0.82	0.88	1.02

In summary, SUNY Geneseo's use of GIST as an entirely patron-driven approach proved to be a beneficial collection development strategy. Acquisitions staff members were able to deliver several thousand items per year within 10 days of a patron's request. When compensating for weekends and holidays, this is generally a turnaround time comparable to Amazon.com's free shipping option of 5-7 business days. Purchase requests were generally more representative of the patron population, as requests by undergraduates outpaced those of teaching faculty—generally in line with the overall request volume received by interlibrary loan. As a result, items purchased in the three years since GIST was implemented circulated 25% more often than those purchased in the previous three years.

GIST has proven to be an effective and efficient system for facilitating patron-driven acquisitions. The results at SUNY Geneseo reinforce the effectiveness of patron-driven acquisitions as a collection development strategy. In an era of decreasing purchasing power, libraries can benefit from using GIST to ensure that they are getting the most for their acquisitions dollars.

Getting started with GIST

If you are interested in implementing GIST at your institution, several resources are available to get you started.

GISTLibrary.org

At http://www.gistlibrary.org, you can download and install the latest version of GIST for ILLiad, including the Acquisitions Manager, which allows you to manage your funds from right within ILLiad. By following the step-by-step documentation, you will be guided through the entire process—from planning to production. Here you will also find the downloadable GIST Workbook, which includes planning activities designed around a team-oriented approach to implementing GIST.

GISTLibrary.org is also the hub for the GIST user community. Users can leave comments throughout the website. Join others in discussing their use of GIST on the GIST-L listserv and community forum. All of these will allow you to share and learn about getting the most from GIST.

Atlas Systems and OCLC support

Sites hosted by OCLC can also take advantage of GIST for ILLiad. OCLC Support will now install the GIST tables to your ILLiad database by request. Simply contact support@oclc.org to get started.

Works cited

Gammon, J., and O'Neill, E. T. 2011. *OhioLINK—OCLC Collection and Circulation Analysis Project 2011*. Retrieved from http://www.oclc.org/research/activities/ohiolink.html.

Kyrillidou, M., and Bland, L. 2009. *ARL Statistics 2007-2008*. Retrieved from http://www.arl.org/bm~doc/arlstat08.pdf.

Mortimore, J. M. 2009. "Access-informed Collection Development and the Academic Library." *Collection Management* 30(3): 21-37. doi: 10.1300/J105v30n03_03

Nisonger, T. E. 2008. "The '80/20 Rule' and Core Journals." *The Serials Librarian* 55(1/2): 62-84. doi:10.1080/03615260801970774

Ward, S. M., Wray, T., and Debus-López, K. E. 2003. "Collection Development Based on Patron Requests: Collaboration Between Interlibrary Loan and Acquisitions." *Library Collections, Acquisitions, & Technical Services* 27: 203-213. doi:10.1016/S1464-9055(03)00051-4

Chapter 2:

Getting It from Start To Finish: The Getting It System Toolkit and Texas Tech University

By Ryan Litsey, Assistant Librarian for Document Delivery, Texas Tech University

Texas Tech University first heard about the Getting It System Toolkit (GIST) for ILLiad web pages when it was first being introduced towards the end of 2010 and the beginning of 2011. We felt we could use the pages to build upon a purchase-on-demand workflow that was implemented a few years back. Our main focus with the implementation of GIST was the desire to provide a more comprehensive set of information for our faculty, staff, and students that would allow them to make an informed decision on how they would like their requests delivered. In 2011 we began to set up the pages and customizations with full implementation in January 2012. Some of our customizations were standard, but we also tried to implement some new ideas that better fit with the Texas Tech model of Document Delivery and see if we could try some different things. The implementation for Texas Tech falls into three categories. The first category focuses on how we integrated the pages into our workflow, including the customizations we made that were unique to Texas Tech. The second category involves the changes we made to the Document Delivery system and workflow to accommodate some of the pages and give us more flexibility. The third category is the overall marketing and branding of Document Delivery that helped us introduce the pages to the University at large.

First, we developed a rather simplistic model to integrate the GIST program into our existing workflow. Texas Tech kept the general layout of the pages, but we sought to include as many of the API connections that we could. We felt that including the API keys that allow input from various pages would help the patron make a more informed decision. One of the key pages we wanted to get information from was WorldCat. Texas Tech is also a member of the Greater Western Library Alliance (GWLA). The GWLA consortium prides itself on very fast turnaround from receipt to delivery. We wanted to highlight that feature by including the GWLA libraries and the expected delivery times using the WorldCat API key. We also included the Amazon and Google books API. Our primary focus with these was to provide patrons with the information to make a sound decision. This decision making process was important to us when we began offering the pages to faculty only. The control of access will be discussed later. Informed decisions decrease the possibility of abuse from random ordering requests.

We also created a series of criteria that linked directly below the request for purchase options to help patrons get an idea of what types of things we would purchase. The purchase criteria are: the book must have been published within the last ten years; the cost of the book must not exceed $100; theses and dissertations will be purchased as hard copy only; textbooks will not be considered; and the books on DVDs must not be part of an ongoing series. While our purchasing criteria might seem stringent, we wanted the Document Delivery purchase-on-demand to supplement, rather than replace, the purchasing done by our subject liaisons.

Even with these criteria we had great success with requests for purchase. From January 2012 through January 2013, there were a total of 358 requests for purchase. This is a very good number considering the criteria might ordinarily rule out many items a faculty member could want. The really interesting statistic is that while there were 358 requests for purchase, we actually sent 318 to acquisitions. This number is telling for many reasons. Not only does it illustrate that faculty are in fact using the system, but, more importantly, it shows that they understand the criteria and an overwhelming majority of requests placed are purchased. In fact, 88% of all requests placed were purchased, which can be considered successful. This high percentage shows that our faculty accepts and understands how the new system works and is taking advantage of the services we offer. Our ultimate goal was to give faculty the resources to make informed decisions. These stats bear the argument out. The faculty has not bombarded the system with a lot of random or unnecessary requests. They have used the system the way we had envisioned and have put the common fears to rest that, once a purchase-on-demand system is unveiled, faculty will no longer contact liaisons for purchases, but rather constantly ask for items from Document Delivery. We demonstrated that, with clear criteria and a good workflow, fears can be mitigated, if not removed all together. Once the requests were placed using the web pages, it was important for us to develop procedures on the client or workflow side to change the process we had already developed.

The developments we made to the existing workflow are the focus of the second category. With the implementation of the GIST web pages, we were able to take advantage of the ILLiad software to modify and streamline our existing purchasing workflow. We felt that our acquisitions department had a good internal workflow for handling purchase requests. However, we realized that, by using the newly implemented ILLiad routing rules and emails, we could enhance the efficient handling of requests for purchase that reached the acquisitions department and keep our subject librarians informed of who had made requests. We accomplished this in a two-fold effort. The first was to develop the ILLiad routing rules so that we could determine from the information included in the request how it should be processed. The information that was key for us included the price, since we have a price limit on our purchasing criteria. Once the routing rules had been set up, we developed a series of emails that could account for the different types of purchase requests. Aside from creating custom routing rules and emails to both acquisitions and the subject librarians, we also implemented an ISI web module that allowed us to control how the pages displayed based on user status. This approach allowed us to complete a tempered rollout of the GIST purchase-on-demand pages. As mentioned earlier, we began with faculty to see how the procedures would work and to iron out any kinks in the process. Once the system was configured with the API keys and the necessary changes to the workflow, we wanted to make sure we made people aware of the service.

The GIST for ILLiad web pages became part of our larger Document Delivery marketing campaign that began in the Spring/Summer of 2012. The marketing campaign included exposure to a variety of services, including flyers, YouTube videos, QR codes, mention of Document Delivery on the libraries walking tour, and a poster series. The marketing campaign allowed us to highlight the changes we had made to our existing service as well as tell the faculty about the purchase-on-demand system. Branding our service allowed us to reintroduce Document Delivery to the university at large as well as change the conventional understanding of what types of services we offer. The marketing campaign was a huge success and it allowed us for the first time to highlight a specific service.

The Getting It System Toolkit has been a welcome addition to the Texas Tech Universities Document Delivery web pages. While we did not make too many cosmetic changes to the appearance of the pages, we did make a few changes on the back end. We customized GIST to stick with the theme of giving the user the best information we could to help him or her make an informed decision. We included the WorldCat API as well as the ones for Amazon and Google books to give the patron an idea of what the book would cost to purchase vs. how long it would take to borrow from another library. Those efforts, combined with our technological changes that streamlined the process from start to finish and expanded access, have helped the GIST implementation at Texas Tech become a rather special success. The marketing was the icing on the cake. It coincided nicely with the debut of this new service to really help brand the department as one of the premier information supply locations on campus.

Chapter 3:

ILLiad Addons at Tompkins Cortland Community College

By Susanna Van Sant, Tompkins Cortland Community College

The library

The Tompkins Cortland Community College Library serves more than 3,000 FTE students in 54 academic programs including business, computers, nursing, media, biotechnology, liberal arts, and criminal justice. The Library aims primarily to support these academic programs, but also strives to meet the leisure reading interests of students, staff, faculty, and the surrounding community. The Library funds its acquisitions through its own budget and with grant funding from Perkins and the Coordinated Collection Development Program of New York State.

The collection

The Library acquires new, primarily English-language materials published in the United States from scholarly as well as general-audience publishers. Primary suppliers are Barnes & Noble and Amazon. The Library has a Corporate Line of Credit and a membership for free, expedited shipping with B&N and our tax exempt status is on file with them. Amazon also has our tax exempt status on file and the Library is an Amazon Prime member. All non-serial acquisitions are firm orders, most paid by credit card; the Library does not utilize either purchase or approval plans. Since August 2010, we have initiated monograph orders using ILLiad and ordered and (copy) cataloged using GIST addons.

The process, step one: ILLiad

There are three main ways the Library builds the regular collection. First, librarians select materials that will support the curriculum and meet known research assignments (i.e., purchased "just in case"). Second, patrons can recommend items for purchase. And, finally, any loan request for a book published within the past five years is considered for potential "purchase on demand" (i.e., bought "just in time").

Librarian liaisons to the five academic program areas identify publications to add to the collection. They search the item in WorldCat or IDS and click our "Get It" ILLiad request button. Librarians log into ILLiad using a fund-specific username (e.g., tc3nurse or tc3sci). Such a logon sets the form defaults to indicate that this request is for purchase, not loan. Librarians add any notes to the request form about special handling, special location or collection code, special supplier, price, preferred format, if any, and submit the request. Requests coming in from these fund logons, with their attending "purchase" flags, are routed directly to a queue called "Liaison Purchase Approved."

Anyone with a TC3 ILLiad account can submit a book request with the option to identify the request as a purchase recommendation instead of a loan. When such purchase suggestions are submitted they are routed to a queue called "Awaiting Liaison Purchase Approval."

All other monograph requests made through ILLiad, whether "Purchase" is selected or not and regardless of logon category or borrower status, are routed to a queue called "Awaiting Request Processing." Requests for locally held items and for assigned textbooks are addressed by circulation and reserves staff. For items not held and published more than five years ago, we first try to borrow. A request for a book published within the past five years, however, is routed to a queue called "Pub Year Purchase Review."

The process, step two: GIST Addons

It is from this point that the ILLiad routing and e-mailing functions, coupled with the GIST Purchase and Cataloging addons that we have implemented, are especially useful and efficient. The three acquisitions streams described above are addressed here in order of complexity.

First, the most decision-making happens with items in the "Pub Year Purchase Review" queue. Using the GIST Purchase Addon Tab, price and availability is checked (based on the ISBN field for Barnes & Noble, and based on what's in the title field for Amazon). Alternate formats are also readily evident on the vendor sites, and a title that is available for prompt delivery (from the vendor, not a third party), costs less than $25, and fits the scope of the collection, is placed in the online shopping cart and ordered immediately. Titles that are available but may be marginal in terms of price or scope may still be purchased if they are held in few SUNY or IDS Project libraries. The ILLiad client enables quick checking of area holdings on the OCLC tab. In this way, the patron will still get the desired materials quickly and at the same time we expand the overall holdings of the system.

If an item is more expensive and out of scope and widely held, we send out the request to be borrowed.

If an item is over this $25 price point but within the scope of the Library's collection, an ILLiad e-mail template is used to ask the librarian for a "Liaison Purchase Review." The librarian can authorize a purchase or instead indicate that we should try to borrow it. If we have already determined that borrowing would be unlikely, or we have tried to borrow but the request has come back unfilled, we instead send an ILLiad e-mail template to the librarian for a "Reference Review," suggesting that the librarian contact the patron to discuss alternate sources.

Once a decision is made to buy, however, we take advantage of the Connexion tab to locate and export the catalog record to Aleph. Here we get the record matching the OCLC number of the request. Occasionally, a search may have to be re-executed to get the best bibliographic record, but unless the original request lacks an OCLC number altogether, there is rarely a need to leave the ILLiad interface to catalog the item.

Second, requests that were initially submitted as purchase recommendations, by using the drop-down options in the request form (see fig. 1), are treated in a similar fashion. They are bought outright unless they are particularly expensive or do not qualify as curricular or leisure reading material. Requests from faculty and staff (recognizable by the format of the username) might be given more leeway with these criteria. The appropriate librarian is notified with an ILLiad e-mail either to approve the purchase, or otherwise follow up with the patron, if the item is not available to buy or is deemed too expensive.

Once orders have been placed for items from either of the above two queues, the request is routed to "Awaiting POD Arrival" to indicate that the item should be held upon arrival and, after shelf-ready processing, the patron is notified.

Finally, least complex are orders from purchase requests that librarians submit. Once ordered and cataloged from within ILLiad, these are routed to the "Awaiting Acquisitions Arrival" queue to indicate that the item should be made available for immediate circulation and not held for a patron. The librarian is sent a notice that the item has arrived. Librarians monitor their orders in the web user interface, not the ILLiad client or Aleph. A librarian's request may be cancelled in ILLiad if it is not yet published, is out of stock, or if there are insufficient funds. Such notes accompany any cancellation. The librarian can later log into ILLiad, review cancelled transactions, and then simply click to resubmit the request at a later time.

Variations

Faculty can also specify if a book request is for an item that should be placed on reserve. These requests are routed to a "Place on Reserve" queue and are usually purchased immediately from the campus bookstore.

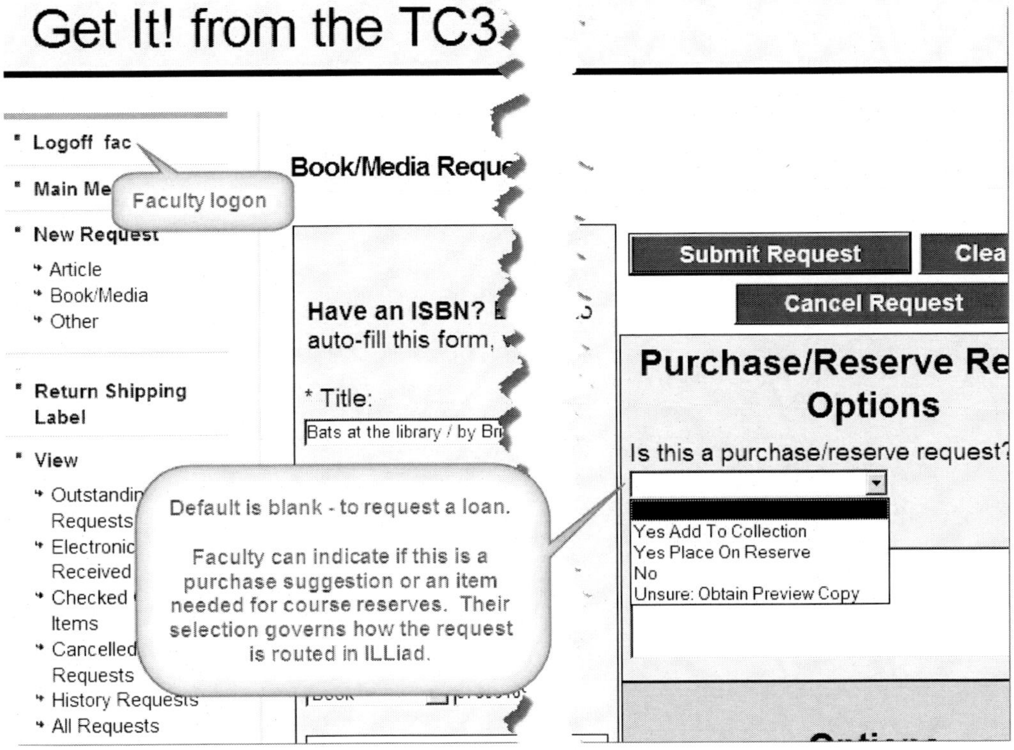

Figure 1. An example of the faculty purchase request interface on TC3.

Considerations: Awaiting request processing

We tried to set up a routing rule based on publication year and local holdings so that borrowing requests for older materials or materials already held would be processed automatically, as usual, and requests for newer and not already owned materials would be diverted from automatic borrowing for our purchase review. However, due to a combination of factors (the sequencing of our other routing rules and inconsistencies in the data in the publication year field), we couldn't set this initial filter reliably. As a result, all monograph requests are now handled manually.

We will monitor the impact on staff to determine if such handling is sustainable or ultimately desirable. This part of the procedure has only been in place for about six weeks, so it's too early to tell. However, it will be easy to turn this setting off over holiday weekends, for example, or if the workflow becomes unmanageable at peak research times of the semester. In the meantime, we are catching requests for current textbooks that are not only unlikely to have been filled, but that enable us to identify gaps in our reserve collection and improve communication with faculty.

Conclusions and Future Plans

Channeling acquisitions through ILLiad and implementing the GIST Purchase and Cataloging Addons have increased efficiencies significantly. Within one application and with only a single sign-on, we can check price and availability, check system holdings, place an order, locate (and export) the catalog record, and indicate handling upon arrival. The time between order request and availability has decreased; internal handling and information-sharing is clarified and documented within ILLiad as needed; and opportunities for communication between liaisons and patrons have been expanded.

Sending e-mails to seek purchase approval from a liaison librarian, while facilitated by the use of e-mail templates, takes us in and out of the client to complete the transaction. However, two of the three liaisons work in the ILLiad client daily, so in many instances a request can be routed to the appropriate queue (or otherwise acted upon) in response to an e-mail request (instead of generating more messages and exchanging information that is recorded neither in the request history nor notes in ILLiad).

We continue to create order records and to process invoices in Aleph. Given the improvements to our workflows and outreach with our current GIST addons, we will be exploring the Acquisitions Addon for a future budget year.

Chapter 4:

GIST Implementation at St. John Fisher College

By Kate Ross, St. John Fisher College, and Micquel Little, Monroe Community College, formerly St. John Fisher College

Subject

St. John Fisher College's Lavery Library implemented GIST (Getting It System Toolkit) with the intention of integrating Purchase on Demand into the Interlibrary Loan process. This strategy was chosen to make the request submission process remain seamless for Lavery patrons while improving our collection through patron-driven acquisitions. Leveraging the power of ILLiad, combined with GIST, we were able to change perceptions about collection development and allow our users to guide the growth of our collection at their curricular, research, or leisure point of need.

Since the initial setup, Lavery has increased the number of items provided to our patrons through ILLiad requests by an average of 22% a year. Our approach continues to be customer driven by focusing efforts on communication between all parties involved, including multiple departments and the patrons themselves. GIST complements this process by allowing us to customize the information gathered, which influences Library decision making and service quality to our patrons.

Background

Lavery Library first implemented GIST in 2009 with the following philosophy: Use GIST to leverage the power of ILLiad to expand input into purchase requests while taking advantage of our simplified Collection Development and Acquisitions processes. GIST implementation coincided with a time of change at Lavery Library. On the heels of Lavery's new membership in the IDS Project, Interlibrary Loan requests were increasing, and library trends nationwide began illustrating the benefits of encouraging faculty, students, and staff to submit library materials purchase requests. Changes in staff and reorganization of departments led to increased collaboration at Lavery as well.

Our implementation began with investigating the GIST development team's proposed workflow. This workflow seemed more complex than Lavery staff desired, so we decided to learn more about best practices for Lavery while moving forward with the implementation of GIST and purchase on demand. Acquisition decisions using GIST were originally centralized through one person within the library. We began with our director piloting this project. At that time, Lavery's director already had experience using the ILLiad client software, so the learning curve was not as steep. In addition, the director's more developed knowledge of the curriculum enabled centralized decision making. It may seem unusual to begin a purchase on demand program with the director in the driver's seat, but doing so at Lavery left the Acquisitions Librarian free to learn how to use ILLiad and examine the processes for continuous quality improvement. Purchase on demand decisions were made within ILLiad at the point of patrons' submitting their Interlibrary Loan book requests. If a loan request was to be purchased rather than borrowed, the information was sent to the Acquisitions department through an email within ILLiad and the request itself was marked as "finished" within the ILLiad workflow. The Technical Services department (an umbrella for both Acquisitions and Cataloging) purchased, received, and cataloged the item. The patron was then notified that their request was available, just as they would have been if the book were borrowed. The process was minimalistic and simple.

Analysis

GIST implementation at Lavery began with a trial and error approach, heavily relying on instincts and flexible improvements. Reflecting on this now, our instincts turned out to be correct: eliminate as many layers as possible from the decision making process and centralize the contact points for all loan requests.

Once the benefits of expanding purchase on demand into existing purchasing workflows were realized, streamlining the purchase on demand process would continue over the next three years. By this point, Lavery's Acquisitions Librarian had a more developed sense of the print collection's strengths and weaknesses and took over the decision making process. The Acquisitions department expanded the use of GIST into the general workflow of ILLiad loan request processing.

Our original goal for implementing the GIST software at Lavery was simply to purchase as many loan requests as met our collection development criteria for purchase. It was our expectation these purchases would generate at least one circulation, which, it could be argued, was better than the more likely zero circulation that the majority of our collection was receiving. This was the approach taken in the initial first year of implementation. During GIST's second year at Lavery, a more careful consideration of the titles being requested was taken and, over time, our goals have been modified to:

- Provide patrons access to what they need, whether it's appropriate to add the title to our collection or borrow it through Interlibrary Loan.
- Communicate with patrons for a good customer service experience to get the best use of library resources.

Using GIST helps us accomplish these goals while taking advantage of our current workflow. More importantly, it also allows us to capitalize on the purchase on demand model without sacrificing our desire to maintain great customer service. Historically, there has been a tension between what users request and titles determined by librarians to be of "value." However, if you know your population and curriculum, your collection development policy can help you fit patrons' actual needs into your purchase decisions; GIST streamlines this process each step of the way. Lavery has learned to leverage the power of ILLiad to simplify Collection Development and Acquisitions processes, while changing perceptions about allowing our users to help develop our collection at their curricular, research, or leisure point of need.

Lavery's workflow begins with analyzing the information patrons provide within the GIST interface on the Interlibrary Loan request forms. The data gathered during request submission are coupled with ILLiad's additional functionality, enabling the completion of the decision making process. The information received helps to inform our decision about whether to add the item to the collection or borrow it. The Acquisitions Librarian evaluates each loan request as a collection development decision. Lavery's priority for deciding what to purchase is not first and foremost about saving money. Each request is evaluated based on the following criteria, in this order: is the request from a faculty member or student? Does the request support a particular course or curriculum? Is it easily available for purchase? The ultimate decision is the result of art, not science, as we have no one-size-fits-all approach. For now, we employ a less-likely to more-likely-to-purchase gradation. While parts of this process can be automated, for example, only reviewing faculty requests, the Acquisitions Librarian is reluctant to do this until request volumes make it absolutely necessary. Using past years' experience of purchasing, we prefer to review each request and recognize its potential to benefit our collection.

Additionally, the GIST software itself brings together helpful tools under the ILLiad umbrella that allow the decision maker to determine whether an item is already owned (Lavery's policy does not include duplication), of value to a college collection (as checked against Resources for College Libraries), a textbook (Lavery's policy states we don't purchase required texts), or freely available as an eBook (via HathiTrust or Google Books). All of these tasks are accomplished using separate resources, but ILLiad brings them all together at your fingertips within each request through its addon feature. Once a decision about the selection is made to purchase or borrow, ILLiad addons can be used to further investigate whether purchasing is still a viable option. In Lavery's workflow, the Acquisitions Librarian determines whether an item is available in stock at Amazon (Lavery enjoys free two day shipping with our Prime subscription) or available to Rush from YBP Library Service's GOBI3 ordering system. The information provided on the user side combines with the addons in the GIST tab to enable the Acquisitions Librarian to quickly decide whether to purchase or borrow. Lavery Library's collaborative approach to GIST requests, with both Acquisitions and Interlibrary Loan departments involved, means requests flow seamlessly for the patron between purchase, borrow, and sometimes back to purchase for additional review if the item is not available to be loaned.

Over time, we began to see that customization of GIST was an important piece of our relentless pursuit of excellent customer service. As mentioned before, feedback and communication from the patrons is so valuable during the Collection Development process. The ability to gather more information from the patrons about their request helps to place the item along the decision continuum. Being able to add/remove GIST fields on the patron request form has enabled us to truly make GIST work for our local purposes.

GIST within ILLiad also enables the features of the software to be utilized and added to the quality customer service experience. Using ILLiad to manage and track communication between Acquisitions and patrons, Acquisitions and the Access Services department, Access Services and the patrons, etc., makes it easy to request more information, send follow up emails, send informational emails, and more during the fulfillment process. Effective communication with patrons does add layers of complexity, but the impact is minimal, as only a small percentage of requests requires more than one communication to make thoughtful collection development decisions. Lavery Library is committed to the idea that its people are its best resource. Therefore, this personal level of feedback regarding a small percentage of requests is not considered inconvenient or unnecessary; rather, we are reinforcing the belief that personal attention is best for our patrons. Our GIST procedure is simply an extension of this personalized attention.

In April 2012, as we increased the extent to which we were communicating with patrons regarding their requests, we were curious about patrons' level of satisfaction with this initiative. We created a short survey and added the link to the notification email patrons receive when we have decided to buy their request instead of borrow. The survey asks, "Regarding your request, how satisfied are you with the communication you received after you submitted the Interlibrary Loan request?" To date we have received 35 results. The results are overwhelmingly positive, with nearly 83% responding that they are Very Satisfied (on a scale of 1-5).

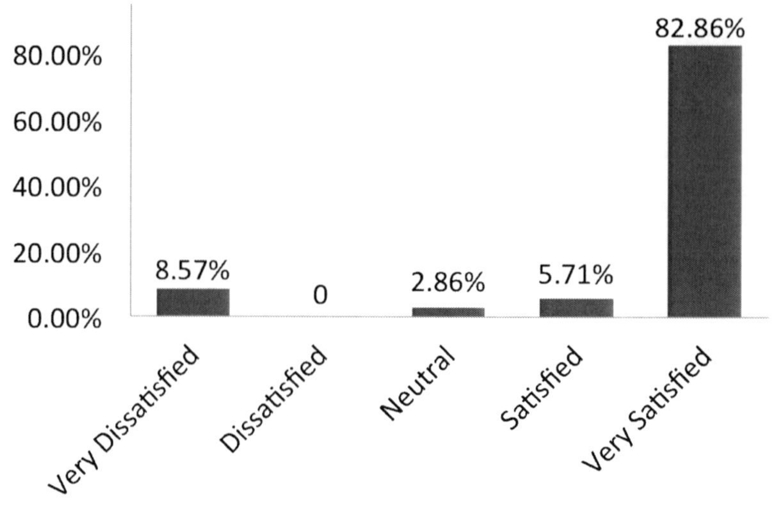

Figure 1. Patron satisfaction with communications about their requests.

Over time, an increased understanding of how GIST works within ILLiad has led to customizations and creative collaborations between library staff to provide the best information specific to Lavery's needs. The daily workflow is still centralized but takes advantage of the more advanced features in ILLiad, enabling a more collaborative approach. Over the last four years of using GIST, Lavery Library added over 1,673 patron-initiated titles to our collection based on collection management decisions. Circulation statistics show that the approach we've taken to purchase on demand encourages decisions that are valuable to our patrons and our collection. Since implementing GIST, the number of Interlibrary Loan requests cancelled has dropped from 24% in academic year 2009/10 to 16% in 2011/12. This has increased the number of books accessed by our patrons to 84%, 27% of those requests being purchased through the use of GIST. Referring back to our circulation statistics, we can see that items purchased using GIST enjoy a much higher percentage of use than items purchased through other collection development methods (i.e., assumption of need, vs. point of need).

Clearly, we continue to meet our original goal to have GIST purchases (on-demand) enjoy at least one circulation. The percentage of our GIST purchases with one circulation has increased over the four years of our program while the percentage of items purchased via traditional collection development methods (in case) with one circulation has remained stable.

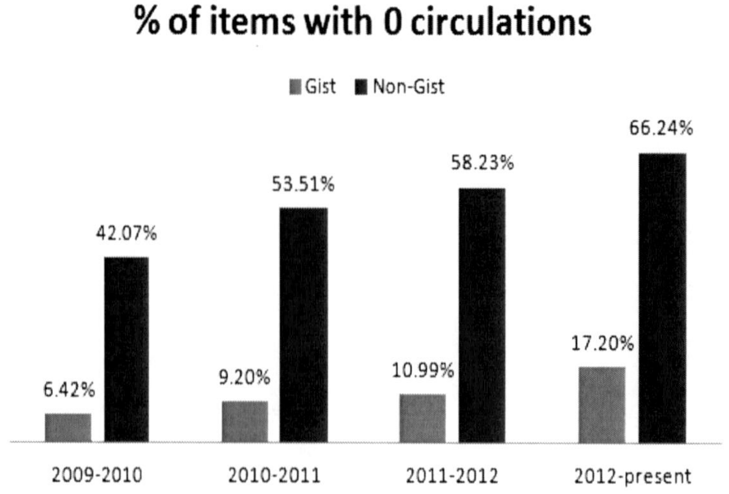

Figure 2. Comparison of zero-circulation items purchased with and without using GIST.

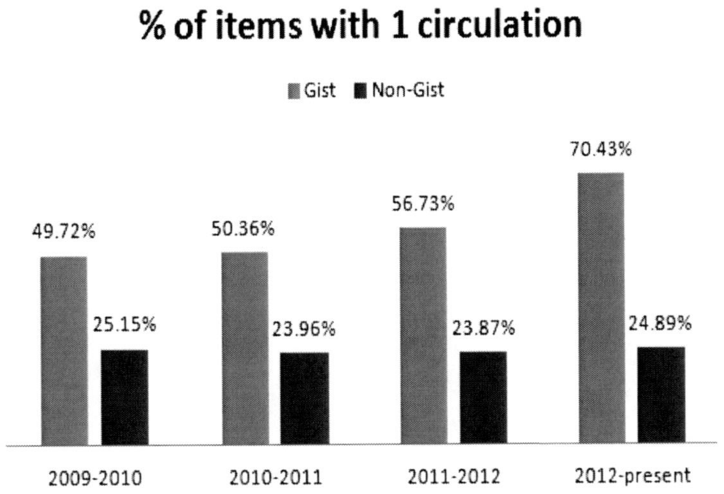

Figure 3. Comparison of single-circulation items purchased with and without using GIST.

Looking beyond the initial circulation of GIST and non-GIST materials, we see that we are moving in the right direction of developing a user driven collection. While the academic library trend is consistently seeing less usage of books as resources by their patrons, the percentage of materials that enjoy more than one usage clarifies our collection development effectiveness. GIST enables us to make this happen.

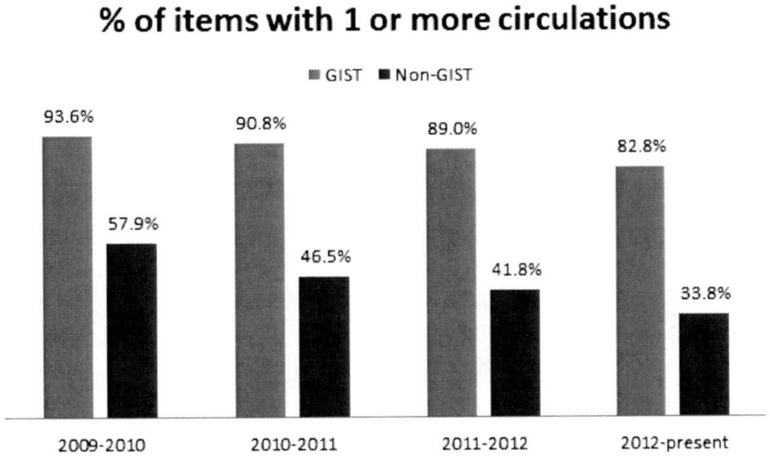

Figure 4. Comparison of multiple-circulation items purchased with and without using GIST.

Lavery Library expanded on the GIST skeleton, adding flesh and muscle to form a complete body of customer service scenarios. We took the available product and enhanced our use to make it work even better within our own environment. We utilized feedback from many participants to create the best, most informed buying decisions for our collection. Ultimately, Lavery Library considers the benefits of meeting the needs of our users while adding items potentially useful to our local population. We weigh this consideration against the costs associated with Interlibrary Loan and the benefit of adding items potentially useful to our resource sharing population.

Cautionary notes

Some library administrators, librarians, and/or staff members may have differing opinions about which department should be responsible for purchasing on-demand requests. GIST purchase requests come in through Interlibrary Loan, though most libraries have a separate department to handle purchasing. At Lavery Library, we kept the GIST purchases in Acquisitions. Our focus on departmental collaboration and utilization of ILLiad allows this process to fit into everyone's existing workflow.

Lavery Library staff could have automated the process significantly; for example, only reviewing faculty requests for purchase. Less attention would have been paid to each request's applicability to our collection and curriculum, however—a sacrifice we have not yet been willing to make.

Chapter 5:

Using GIST in Response to User Feedback

By Kerri Goergen-Doll, Oregon State University

Many libraries that have undertaken a purchase-on-demand program have written on their implementation process and have evaluated the service through a return on investment (ROI) lens based on circulation of material. A book purchased and checked out a specified number of times is a sound financial investment. Oregon State University Libraries (OSUL) confirmed this with the evaluation of their pilot service, Buy Request. Their evaluation also uncovered a missed opportunity to include the expertise of faculty and graduate students in the materials selection process (Hussong-Christian and Goergen-Doll 2010). Scholars submitting requests indicated they wanted to convey their preference for what titles should be considered for purchase. OSUL's existing process for turning ILL loan requests into purchases did not utilize this valuable feedback from discipline experts.

As OSUL moved to make the Buy Request service permanent in 2010, two options to incorporate scholar feedback were considered: either repurpose ILLiad fields on the existing ILL web form, or install the GIST web form. The GIST web form was selected because it not only allows purchase suggestions from requestors, but it also provides information that can help assist in their decision making process. Respondents to the Buy Request pilot evaluation survey often noted an understanding of library budget constraints (Hussong-Christian and Goergen-Doll 2010). The GIST APIs that display estimated title costs from vendors (Amazon and Better World Books) and delivery times through resource sharing channels provide scholars requesting material with valuable information and help them balance their needs against OSUL's budgetary constraints.

Scholars' feedback

The GIST web form was launched at OSUL in December 2011. A survey was sent to patrons submitting loan requests between February and April 2012 (12 weeks total). During that time 146 of the 455 eligible survey respondents completed the survey on their experience requesting a loan using the GIST web form. Overall, response to the new form was positive with 86% of respondents indicating the form was easy to use, and 89% indicating the information provided on the form was useful.

Survey respondents were specifically asked if they used the "feedback" section of the GIST web form to recommend a book purchase. Response options included "yes," "no," and "didn't see it." The "feedback" section was not seen by 36% of respondents, and was not used by another 46%. Based on the results of this survey, and the one conducted during the pilot evaluation of Buy Request, it is estimated that the population of requestors interested in providing their input to purchasing decisions is small but vocal. This group consists of faculty and graduates that are moderate to heavy users of ILL services.

The GIST API that displays the price of titles from Amazon and Better World Books was not seen by 63% of respondents. One responder to the survey commented that s/he thought the Amazon price was put there to encourage requestors to buy it for themselves which s/he thought odd of a library. That so many respondents did not see this API in action could be an indication that the requestors lacked an ISBN, a requirement for the API to display this information. This would be the case for requestors who manually entered citation information into the GIST web form since use of a link resolver would populate the form with an ISBN.

In looking at the dollars and "sense" of using GIST, the typical ROI of just-in-time acquisitions needs to be evaluated. During the time of the GIST web form survey (February–April 2012), if OSUL had purchased every loan requested through ILL that was available from Amazon (the Buy Request vendor) the total cost would have been $67,959.16 (not including shipping). If OSUL bought everything indicated for purchase by scholars on the GIST web form, the cost would have been $8,563.34. Using only the criteria established for Buy Request, expenditures during February through April 2012 totaled $878.17. During the same time period, OSUL actually purchased $117,212.40 in firm orders and through approval plans. The circulation of material purchased just in time through the pilot phase of Buy Request was 215% greater than that of firm and approval orders (1.6 circulations compared to 0.5 circulations) (Hussong-Christian and Goergen-Doll 2010).

Although the overall price is lower when not using scholars' feedback, the established Buy Request criteria are limited to concrete factors of cost and format. The criteria do not take into consideration factors such as intended use and impact of the title on current and future research, nor do they include the scholar's subject expertise. Unless a librarian is embedded in the research being conducted in the different areas on campus, s/he is also unable to predict intended use and impact for every single title.

Data-driven decisions

Although the GIST web form was launched in December 2011, purchase suggestions submitted by requestors were not incorporated into the Buy Request criteria until January 2013. This allowed for the collection and analysis of the requests that helped inform decisions on what criterion to adjust for the Buy Request program.

ILLiad was queried for loans submitted between December 2011 and November 2012. Of the 6,076 loan requests submitted during that time, 20% of the requests included input on purchasing titles. Input options available to requestors are "purchase," "do not purchase," and "unsure."

Concerns with allowing requestors to suggest titles for purchase included fears that loan requests would increase overall, and that undergraduates would monopolize the service with requests, particularly for textbooks. The number of requests submitted using the GIST web form was slightly less than the average of the two previous years (6,710). Of requests that indicated the title should be purchased, only 26% were submitted by undergraduates. Purchase requests for textbooks from any status of user (undergraduate, faculty, graduate, etc.) are not fulfilled because current criteria prohibit purchasing textbooks.

Prices could be found for 362 of the titles suggested for purchase using Amazon or GOBI. The grand total was $36,722.18 with the least expensive title at $8.95 and the most expensive title at $2,551. If titles that were available through Summit libraries were removed, the total cost dropped to $28,060.32. Summit is the resource sharing component of the Orbis Cascade Alliance (OCA) of which OSUL is a member. Only 18 of the 362 titles were over $250, and 12 of those were available from Summit or another reciprocal consortium to which OSUL belongs.

As part of OCA, OSUL participates with the other member libraries in a variety of endeavors including collection development and resource sharing (Summit). A voluntary collection development agreement for member libraries is to avoid adding a fourth copy of a title. Of titles requested for purchase, 73% were not available from a participating Summit library. Because GIST at OSUL is configured with the OCLC symbols of libraries that participate in Summit, it is very easy for staff to see the overall holdings of the consortium and process the requests to borrow accordingly.

After reviewing the analysis of the GIST web form requests with the OSUL collection development team, the maximum cost for purchasing a title through Buy Request was raised from $150 to $250. Buy Request was also opened up to every status on campus (faculty, staff, graduate, and undergraduate) that is eligible for ILL services. EBooks will now be purchased when available through established vendors. The previous format criterion was limited to print only. This analysis also confirmed that the voluntary collection development agreement for titles already held by three or more OCA libraries could easily be managed using GIST.

Impact potential

OSUL has moved to e-preferred format for content selected for the collection in general. This allows immediate access to OSU scholars, especially those at a distance from the physical library. The GIST web form allows requestors to inform OSUL of their preference for print or electronic books. Understanding format preference allows opportunities for targeted action. Followup with individuals who did not want to accept electronic books can be conducted in order to determine barriers. Any barriers perceived by requestors can be collected and shared with vendors as a means to improve the overall usability of e-content.

The GIST web form also offers the opportunity to harvest data pertaining to OSUL's impact on research and teaching. If a title is purchased because a scholar indicated it was essential to their research or classroom use, followup with how the material was used is very easy. Directly tying the act of purchasing titles for the collection with the research and teaching of OSU scholars continues to enforce the importance of the library in helping meet the overall mission of the university. The ease of customizing the GIST web format allows any institution the ability to select impact or importance factors that need to be demonstrated.

The GIST implementation has allowed OSUL to incorporate the expertise of OSU scholars into the process of enhancing the collection, to make data-driven decisions, and to see opportunities to demonstrate the impact of OSUL on the teaching and research mission of the university. Not only does this increase the potential ROI of titles added to the collection, it also builds relationships between OSUL and OSU scholars. OSUL looks forward to investigating all the GIST tools available from the IDS Project, and finding how the tools can be used to improve service and demonstrate impact at OSUL.

Works cited

Hussong-Christian, Uta and Kerri Goergen-Doll. 2010. "We're Listening: Using Patron Feedback to Assess and Enhance Purchase on Demand." *Journal of Interlibrary Loan, Document Delivery & Electronic Reserve* 20.5: 319–335. Print.

Chapter 6:

The Use of GIST at Old Dominion University

By Katherine Mason, Central Michigan University, Park Library, formerly at Old Dominion University, Perry Library

The path to implementing the Getting It System Toolkit (GIST) at Old Dominion University has been a long and winding road with unplanned obstacles and setbacks. As of May 2013, the group formed to implement GIST suspended their activities due to ongoing issues with adding data to OCLC servers. Since implementation was not completed, this paper will focus mostly upon the exploration and planning process leading up to GIST implementation.

GIST came to the attention of librarians at Old Dominion University in spring of 2011. The Information Delivery Services (IDS) Librarian stopped at the IDS Project table at the annual ILLiad International Conference and was intrigued by the possibility of using ILLiad interlibrary loan software, an existing tool, to facilitate and improve communication between the interlibrary loan and collection development units. Around the same time, one of ODU's Associate University Librarians forwarded Kate Pitcher's February 2011 Against the Grain article, "The Getting It System Toolkit (GIST) and Changing Workflow in Acquisitions and Collection Development," to a small group of librarians from resource sharing, acquisitions, bibliographic services, and collection development, suggesting that the group explore the possibility of GIST at ODU.

In the summer of 2011 a small discussion group was formed to explore GIST and determine the feasibility of implementing the system at ODU. In addition to the Against the Grain article, the group conducted a brief literature search and contacted the IDS Project for more information about GIST. Two group members also attended a presentation by T. Jacob Weiner of George Mason University about GIST at the Virtual Library of Virginia (VIVA) ILL Forum in July 2011 and followed up with questions via email. The GIST discussion group met three times to review the available information, to determine what resources would be needed to implement GIST, and to decide whether or not to recommend that the project proceed.

In September 2011, the group submitted its report to the library administration team with a recommendation that Old Dominion University Libraries move forward with GIST implementation due to the many possible advantages and minimal risk or cost to the Library. The possible advantages identified included:

- More efficient workload though automation
- Improved communication
- Interoperability with existing systems (e.g., ILLiad, GOBI, WorldCat, etc.)
- Customizable to the library's needs
- Excellent statistical and record keeping functionality

While waiting for a decision from library administration, the Acquisitions Librarian and IDS Librarian developed a list of nine tasks required to implement GIST: select GIST components that meet the library's needs; plan and set up the workflows; develop criteria for selecting requests for acquisitions; customize ILLiad web pages; set up ILLiad accounts for acquisitions and collection development staff who would be using GIST; install ILLiad and GIST addons on all GIST participants computers; determine how project would be funded (GIST-only fund or from bibliographer funds or other fund); train staff; and develop criteria to evaluate GIST activity.

During this same period, Atlas Systems, Inc. announced that they were starting a service to provide training assistance to libraries interested in implementing GIST. The IDS Librarian contacted Atlas about the new service and learned that they were just starting to develop their training program for GIST and were not yet offering the service; however, since ODU and Atlas are located in neighboring cities in southeastern Virginia, Atlas offered to assist ODU with GIST implementation as a test case for their training program. This offer was added to the recommendation and implementation needs submitted to library administration and within a month the GIST pilot project was approved.

At this time, the GIST group at ODU expanded to include a System Specialist from Systems Development, the Heads of Access and Bibliographic Services, and additional staff from Acquisitions who would be GIST users. The first joint meeting of the ODU group and representatives from Atlas Systems, Inc. occurred in December 2011 with a followup meeting planned for early January 2012.

The planning process was systematic and thorough; for each element of GIST, the ODU group selected the parts they would use, workflows, and decision making criteria. Acquisitions staff selected which addons would be most useful in their decision making and determined the criteria they would use to proceed with a purchase or return a request for interlibrary loan processing. The group developed criteria to be used to route requests to collection development, both automatically at the time of submission and manually from interlibrary loan, and created ILLiad email templates to communicate with subject bibliographers and with library patrons.

In addition to planning meetings, the members of the ODU GIST group met to share and learn; in order to understand better the operations of acquisitions and interlibrary loan, staff from both units demonstrated their software and some of the general tasks they perform in the course of their work. This activity helped inform decision making and provided all group participants with greater awareness of library process and procedure.

The ODU GIST group met with and without Atlas Systems several times during January and February of 2012; the initial timeline for implementation was March 2012 and significant progress was being made towards meeting this goal. Each joint meeting ended with "homework" for the ODU group, from making decisions about the criteria described above to collating the local catalog's search string, listing OCLC symbols for a library group availability field on the web pages, and obtaining API keys from WorldCat and Amazon. These activities were mostly straightforward, though acquiring API keys from Amazon required multiple attempts and the Amazon widget for the web page would not display correctly.

As Atlas was preparing to schedule their on-site visit to ODU Libraries, they discovered a problem with loading the GIST tables into the ILLiad; ODU is a hosted site and their ILLiad data are stored on OCLC servers. OCLC did not have a procedure or policy for accepting third party data on their servers, so progress on the GIST pilot project came to a halt while OCLC developed a process for vetting and accepting third party data. This process took several months and in mid-December 2012, ODU received word that OCLC was prepared to accept the third party data for GIST implementation.

In early 2013, the ODU GIST group and Atlas Systems, Inc. representatives met again to regroup and identify final steps to implementation. The ODU group members reviewed and reaffirmed their decisions made earlier in the process. At the time of this writing the Amazon element of GIST does not function properly, though both the ODU group and Atlas are considering alternatives to work around this issue if necessary.

The GIST pilot project has not been as straightforward and simple as the GIST discussion group envisioned, yet the project has been meaningful and enjoyable. Group participants increased their knowledge of library operations and the discussions led to more thoughtful activities between interlibrary loan, acquisitions, and collection development. Despite the obstacles, GIST has been a worthwhile endeavor, sparking conversation and encouraging improved communications. Further, the assistance provided by Atlas Systems, Inc. has been invaluable and their willingness to work with OCLC to allow for external data has allowed ODU Libraries to move forward with implementation, which should be complete in 2013.

Works cited

Pitcher, Kate. 2011. "The Getting It System Toolkit (GIST) and Changing Workflow in Acquisitions and Collection Development." *Against the Grain* 23.1: 73-77. Web. 8 Feb. 2009.

GIST Gift and Deselection Manager

Chapter 7:

GIST Gift & Deselection Manager (GDM)

By Kate Pitcher, Head of Technical Services & Collection Development, SUNY Geneseo

Increasingly, many library operations are moving away from maintenance and support of overly complex processes and systems to actively engage and participate in new roles on the frontiers of publishing, online learning, and digital scholarship. It is crucial for these libraries, and more specifically, their Technical Services units, to begin making important changes in workflow. As library staff members engage in new roles, there is less time for managing processes and workflows that consume inordinate amounts of time. An example is the marked increase in libraries weeding aging collections—although this needs to done to make room for new materials, services, and learning spaces—many libraries are overwhelmed about how to make effective decisions that reflect their thoughtfulness and careful attention to detail and also speed up the process in order to focus time and energy on new responsibilities. The same dilemma arises with gifts and donations. While libraries want to devote more time to actively managing relationships (rather than collections or processes), they need tools to make their workflows and processes automated, streamlined, and smart. We want the donors' goodwill, but we also don't want to let books and gifts pile up in a back room, waiting for a librarian to manually evaluate each and every item before making a careful decision. The GIST Gift and Deselection Manager is designed to ameliorate both problems and automate much of the decision making that takes up too much time in a library's workflow.

Gift processing with the GIST Gift and Deselection Manager (GDM)

Technical Services librarians and staff know too well the problem presented by gifts and donations. Gifts may pile up in back workrooms until the library has space, time, staffing, and resources to make a careful decision about whether or not to refuse, discard, or keep a donated item. In many cases, libraries actively refuse and turn away hundreds of donations a year because of the staffing and time involved to process them. This is unfortunate, as many of these gifts can add value to collections when kept, or make a small profit when sent to a third party bookseller such as Better World Books. A commitment from the library is necessary to change workflow, streamline decision making, and automate functions traditionally handled manually.

The GIST Gift and Deselection Manager is part of the Getting It System Toolkit project and is focused on optimizing workflow by automating most of the work. It leverages systems to provide data for quickly making decisions and shortens the time needed for processing. GIST GDM does this as free, standalone software that is easy to customize and use. It automates the evaluation process, cataloging options, decision making, and gifts and deselection workflow, with immediate results and recommendations for further action (see fig 1).

Figure 1. Item information returned from search, with recommendation action circled.

Because GDM automates the gathering of data for evaluating donations, it helps streamline workflow by:

- recommending *Keep* or *Do not keep* according to your collection building Conspectus profile;
- cataloging the item from GDM by opening and retrieving the record in OCLC Connexion (as you process gifts), to attach your holdings and begin the process to import into your library system; and
- automating the printing of donor letters with lists of donated items (data now stored in GDM's database).

Using and navigating the GIST GDM interface

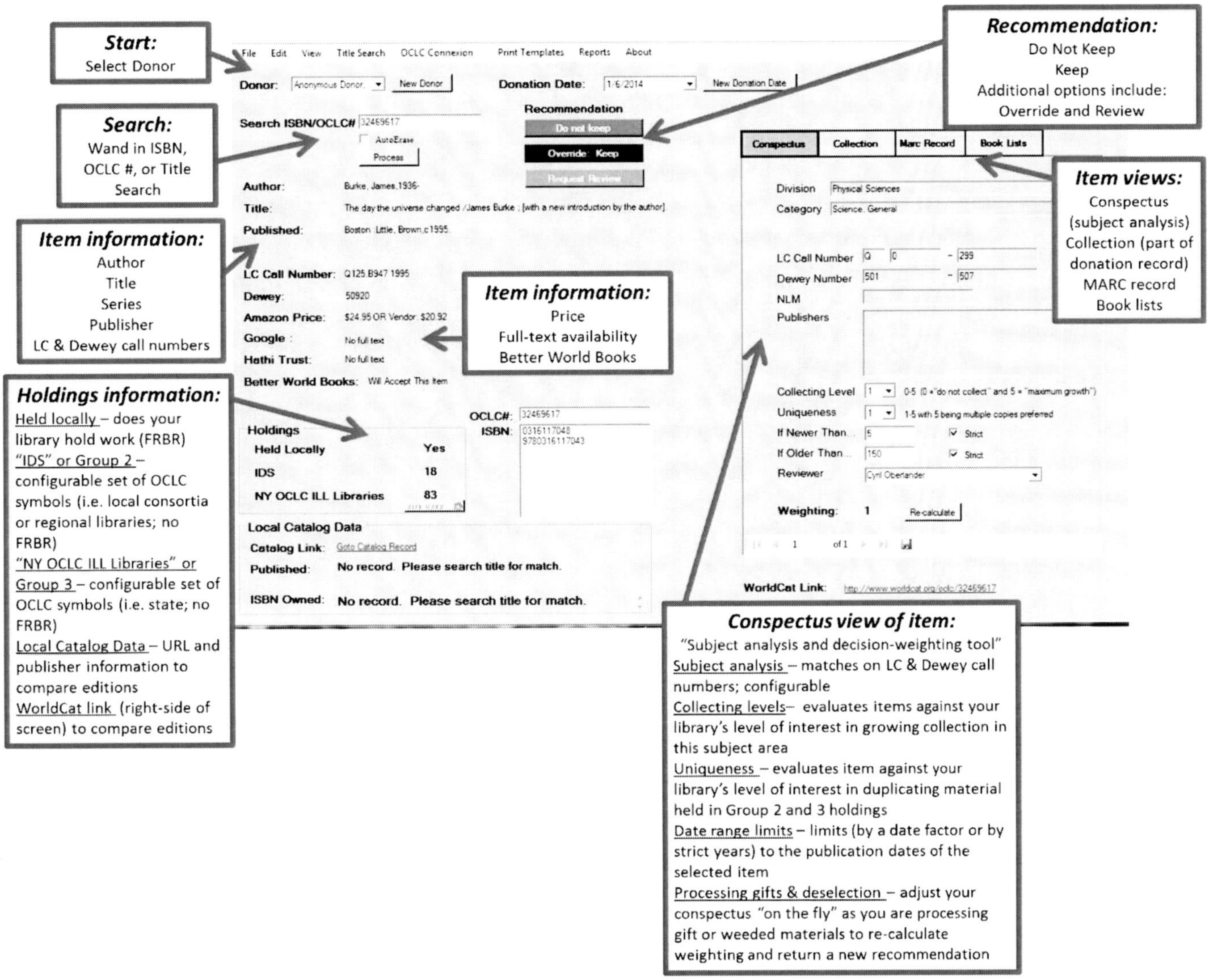

Figure 2. Title-level view of the GIST Gift and Deselection Manager interface.

The GDM interface is designed to make decision making as easy as typing or wanding in an ISBN and letting the system recommend the next step based on your collection building profile.

Steps

Start by selecting a donor—"anonymous" may be the most common, especially if you have a donation box in your library.

Select a donation date. You can select a previous date if you are processing a prior day or days' donations.

Next, either wand in a book's ISBN, type in the OCLC# or do a title search, and press Process (see fig. 3). GDM calls out to the WorldCat API and other APIs to gather all the data for evaluation. The WorldCat API is a call out to the WorldCat database for holdings information, based on the profile you originally set up in the GDM configuration.

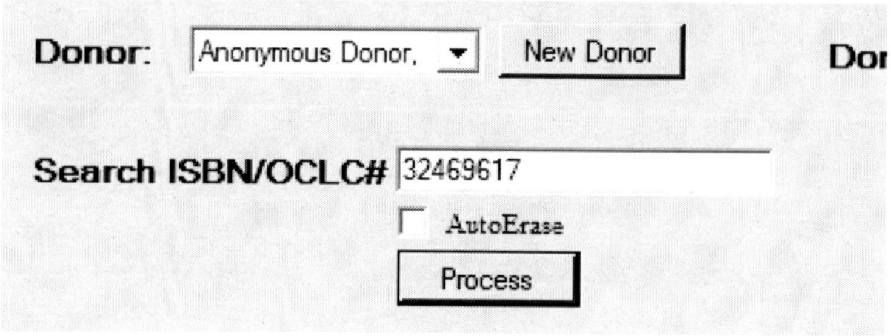

Figure 3. Donor selection and search box for ISBN or OCLC number.

Item information is returned that shows the author, title, series information, imprint, and Library of Congress Dewey call numbers. The price of the book in Amazon (if available) is included, with a link to view more, and full-text indicators at Google Books and HathiTrust are shown, with a link to view more directly. Lastly, the OCLC # and related ISBNs appear (see fig. 4).

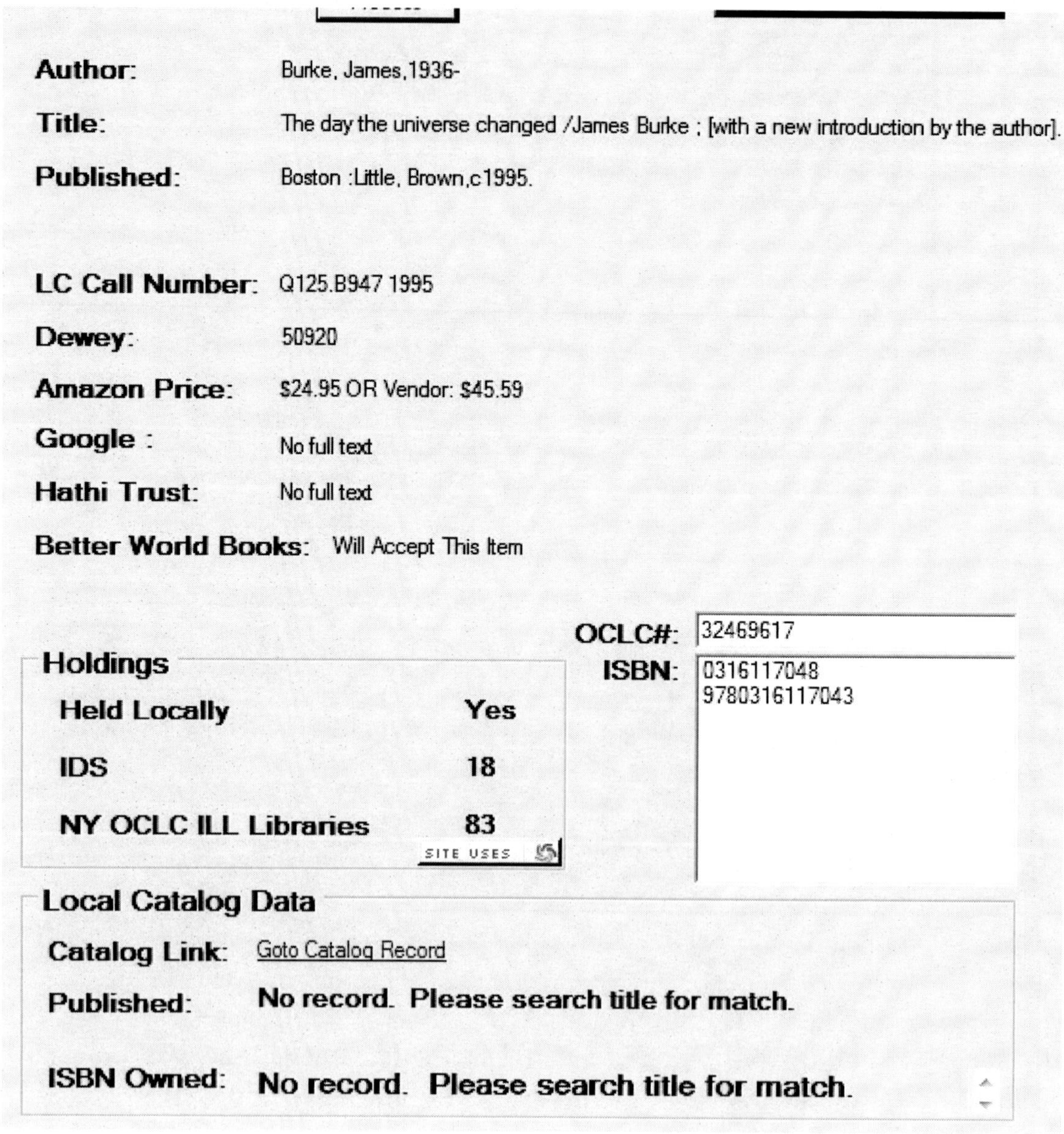

Figure 4. Item information returned from search.

Below the bibliographic information are the holdings data. The first information set is returned from OCLC and provides a FRBR lookup of your holdings and two other customizable groups. The next information set is returned from a lookup of your local catalog, showing a link to your item information and imprint data (for edition comparison check). It also indicates which ISBN you own (see fig. 5).

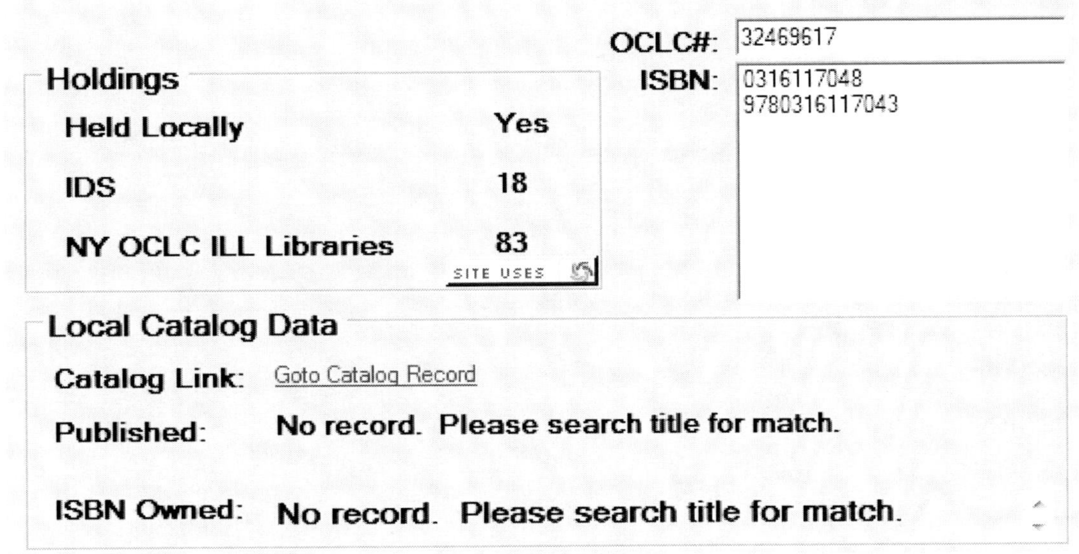

Figure 5. Holdings information returned from search.

The Conspectus view is a collection building profile matching the best LC call number or Dewey classification using the OCLC Conspectus, which you customize. A weighting formula is calculated to make *Keep* or *Do not keep* recommendations based on simple values; how much do duplication, publication date, and growth in this subject area matter? The Conspectus also provides you with the ability to add reviewers if you want to get input from other librarians.

The GDM results and Conspectus provide a recommendation; however, the *Request Review* process allows the staff to send items for review (to a reviewer previously assigned in that subject area) through email or a print slip with the book. Although GDM will recommend an action based on your collection building Conspectus, you can override the recommendation at any time.

As you view information about the item, three tabs on the right side of the GDM interface provide even more information. First, the *Conspectus* tab lists the customized Conspectus information for this item; the *Collection* tab lists items you have been processing for this donor; and the *MARC record* tab gives a detailed view of the record. Currently, GDM is configured with the Caldecott and Newbery Awards and Honor book lists. If libraries are interested, you can upload your own customized book lists for flagging when processing gifts. For example, you can create a book list of wish list items for your library—such as high demand, or high circulating items that you may not want to purchase, but would eagerly accept as a duplicate. You can download a template for creating lists on the GIST website. A *Book Lists* tab will appear and flag items with a red tab when an item appears on a designated award winner book list that you have previously uploaded (see fig. 6).

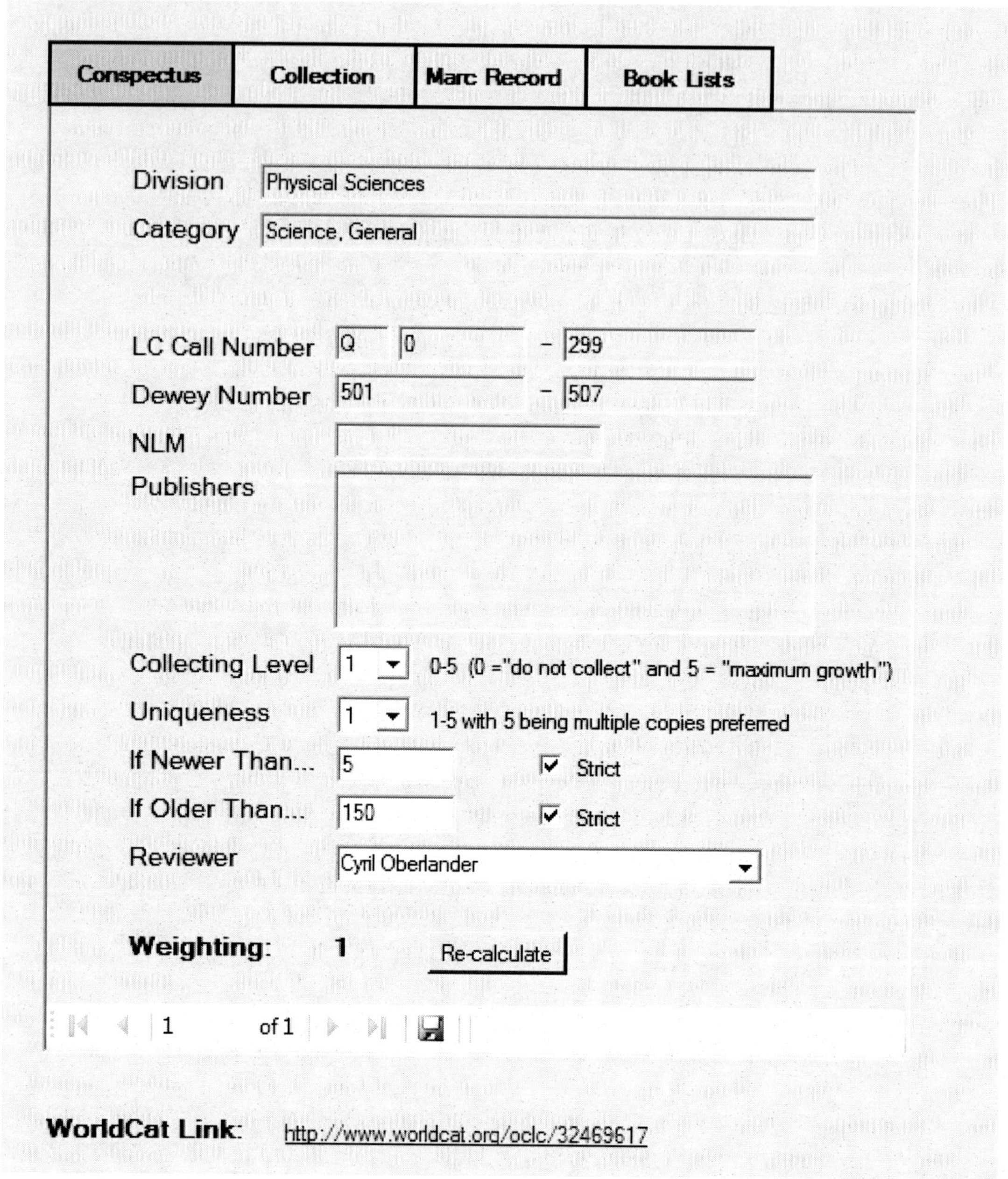

Figure 6. Results screen returns information about Conspectus, Collection, MARC record and Book List if applicable.

After processing all the donated items for a particular donor, click on *Print Template* and select *Thank you Letter*—GDM does the rest, adding all the titles and donor information to a word template—ready to print. The template is found within the GDM software and can be customized by your library staff.

All documentation and installation instructions are found at http://www.gistlibrary.org/ .

Customizing the Conspectus and automating your collection building profile in GIST GDM

The GIST GDM uses the OCLC Conspectus, a subject hierarchy consisting of divisions, categories, and subjects. The Conspectus provides a framework to describe library collections, and is mapped to Dewey Decimal, Library of Congress, and National Library of Medicine classification schemes.

The GIST GDM uses only the Dewey Decimal and LC classification schemes. The Conspectus is structured in hierarchical order, from broad divisions to very specific subjects, and each level is increasingly more detailed.

Configuring the Conspectus

Setting up the Conspectus in GDM allows libraries to customize and automate the gift management workflow. Libraries may determine which areas of the Conspectus they might want to collect and which other subject areas are out of scope. Making these decisions ahead of time and customizing GDM will automate much of the evaluation workflow. Libraries can decide what divisions and categories they wish to grow and then set up collecting levels and uniqueness factors to automate those decisions. Libraries can also set up strict guidelines for keeping or not keeping publications dependent on date limits. All of this customization may be set up before you start processing or evaluating gifts and deselected materials by using the Conspectus editor available from the taskbar under View.

For example, if you wanted to collect comprehensively in Geography and Earth Sciences—Meteorology, but only within the last five years of publication, you would want to set your collecting level to "5," a uniqueness factor of "1" (to avoid duplication in your collection), and set "If Newer than" settings to 5 and click in the box for Strict (see fig. 7).

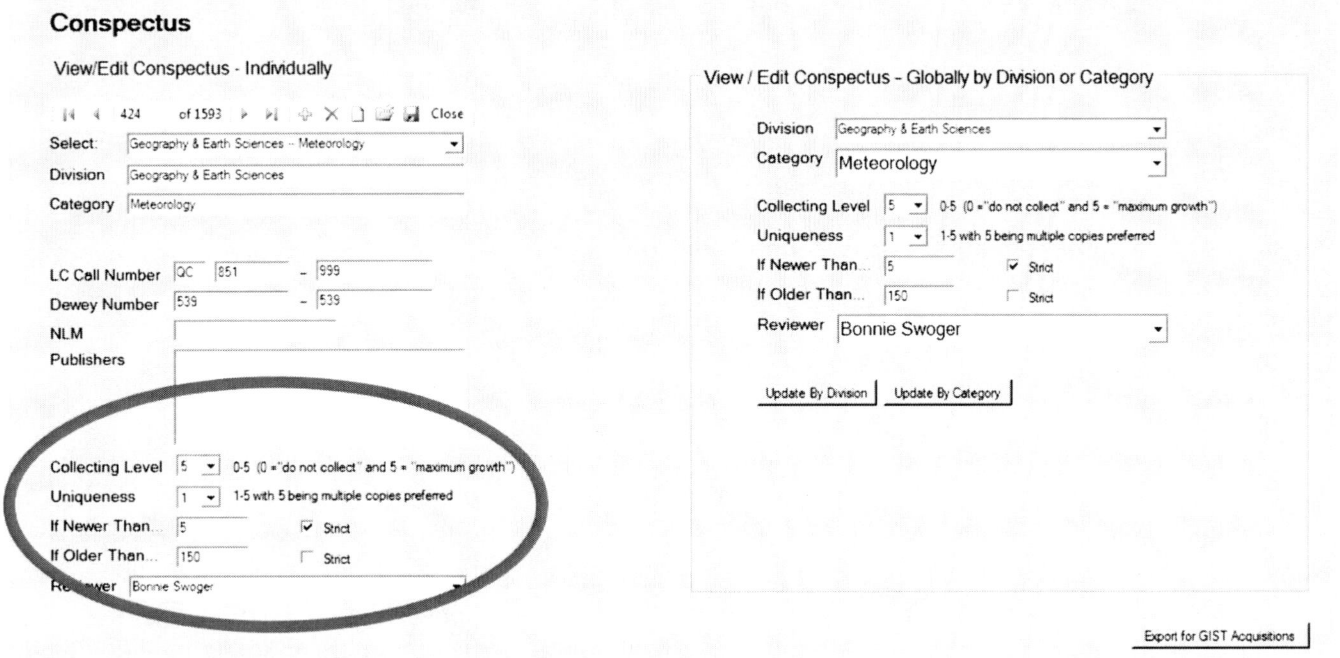

Figure 7. Use the Conspectus editor to set up a customizable, collection-building profile.

Publishers

Use this text area to define publishers that you target for a specific category. While reviewing an item in the GDM, this will allow staff to easily know whether or not it is worth keeping, based upon whether that publisher is preferred or not.

Collecting Level

Collecting Level indicates the depth of your collection-building, with 0 indicating no growth and 5 indicating a comprehensive collection, with maximum growth:

0 = OUT OF SCOPE—Library does not collect in this subject.

1 = MINIMAL INFORMATION LEVEL—Collections that support minimal inquiries about this subject and include a very limited collection of general resources, including monographs and reference works.

2 = BASIC INFORMATION LEVEL—Collections that introduce and define a subject, indicate the varieties of information available elsewhere, and support the needs of general library users through the first two years of college instruction.

3 = STUDY OR INSTRUCTIONAL SUPPORT LEVEL—Collections that provide information about a subject in a systematic way, but at a level of less than research intensity, and support the needs of general library users through college and beginning graduate instruction.

4 = RESEARCH LEVEL—Collections that contain the major published source materials required for doctoral study and independent research.

5 = COMPREHENSIVE LEVEL—Collections in a specifically defined field of knowledge that strive to be exhaustive as far as is reasonably possible (i.e., "special collections").

Uniqueness

The uniqueness factor indicates your willingness to have duplicate copies of materials within this category or division, with 0 indicating no duplication and 5 indicating liberal duplication. This will have a lot to do with the average usage of items in a given subject area at your institution. Take into consideration whether a subject has commonly assigned textbooks, is a specialization of your institution, or has high circulation volume. Remember that this value is distinct from your group holdings and specifically refers to duplicate copies within your own local collection.

Publication dates

You can specify collection limits for a subject based upon the age (in years) of an item. Select the "If Newer Than" preference for items published in the past x years (where x indicates the number of years you want as your threshold); the "If Older Than" preference for items that were published over x years ago; and "Strict" means that no items can exceed the values you entered.

For example: you prefer items published within the past five years, but will still consider items older than that. Set "If Newer Than" to 5. The result would be a higher weighting of items published within the past five years, with older items receiving a diminished weighting. But, if you wanted only items published within the past five years, you would select the "Strict" check box for the "If Newer Than" value.

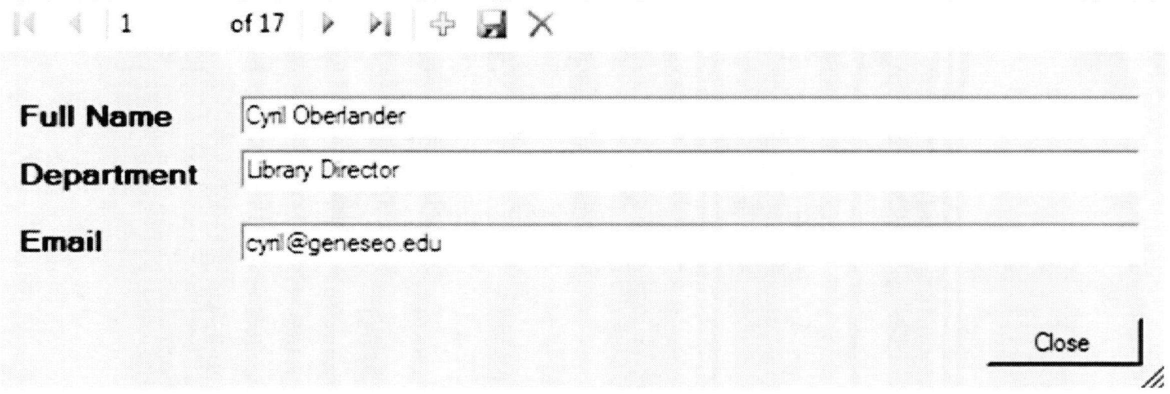

Figure 8. Use the reviewer editor to set up contact information for your faculty and librarian reviewers.

Reviewers

For each division and category, you can specify a reviewer. This is useful if you want to defer a decision to a subject specialist or department liaison. This person will receive an email notification sent from within the GDM software, notifying the person that an item requires review. To create a new reviewer, return to the GDM window and click on the View menu (see fig. 8). Select *Reviewer.* To view in the *Reviewer* window, press the + button to add. Remember to click the save button when you're finished. To add a *Reviewer* to a subject area, you may assign particular reviewers to an overall division level or specifically to categories within that division by using the Conspectus editor and then clicking on *Update Category* or *Update Division* within the category's record (see fig. 9).

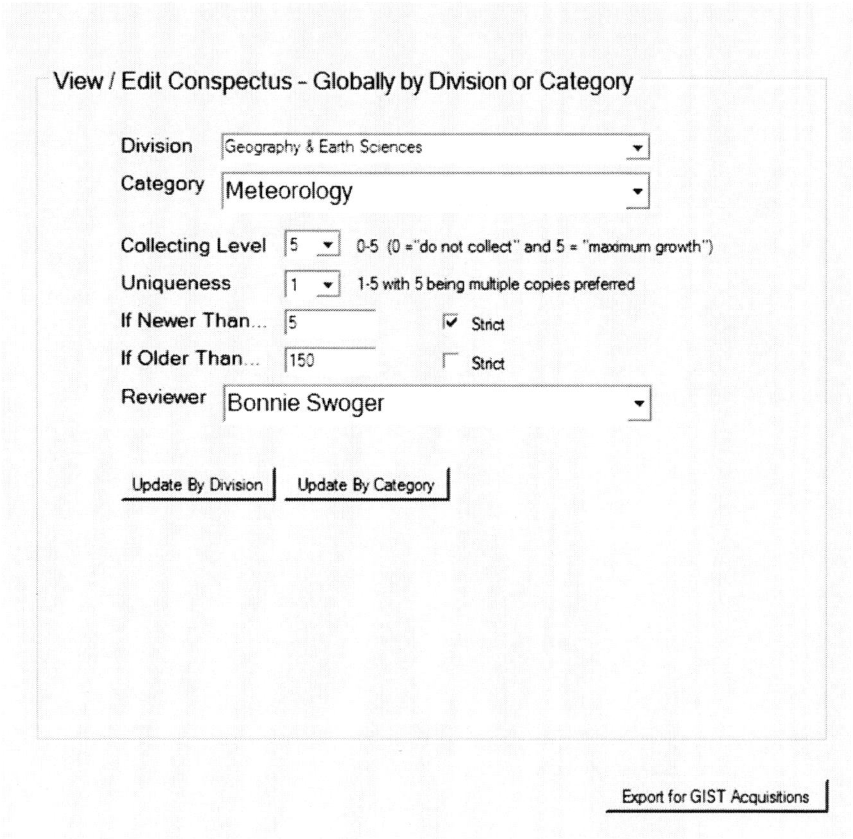

Figure 9. Use the Conspectus editor to update your conspectus by either division or category.

Weighting

The weight score is another easy way to quickly determine the strength of an item's value, in regard to keeping or removing from your collection. In item-by-item analysis, the final weight score will be found at the bottom of the *Conspectus* view on the results screen (see fig. 10).

A number of factors comprise the *Weight* score to effectively determine whether or not GDM returns a *Keep* or *Do Not Keep* recommendation. Holdings, publication dates, uniqueness, and collecting level are factors affecting the final weight of an item. A higher weight score gives a recommendation to keep an item, due to these factors. A lower weight indicates an item of less desirability than the highly weighted item (see fig. 10). For example, at Geneseo we have set our weighting score threshold at 100. This means items with a weighting score of over 100 are items to *Keep*, while scores of under 100 we automatically flag as *Do Not Keep*.

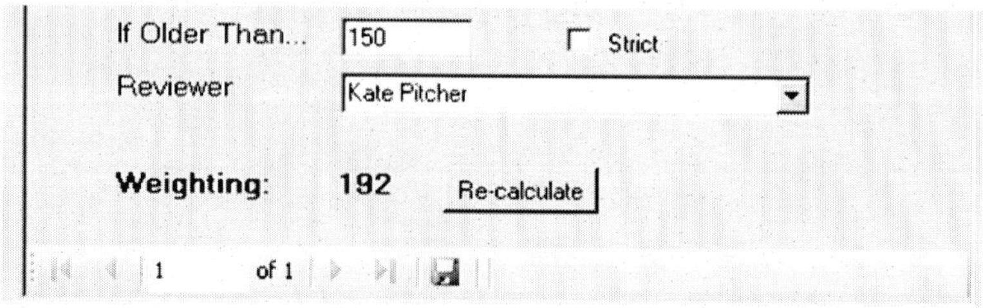

Figure 10. Weighting score from an item analysis in Deselection Manager mode.

The formula that is used to calculate the score is a configurable number and the first step of the logic is to determine if either of the strict date restrictions applies.

- If "Newer Than Strict" is checked and the item is older than the limit; or if "Older Than Strict" is checked and the item is older than the limit, then the preliminary weight (*pWeight*) **is automatically set to 500**.
- Otherwise, the *pWeight* is based on the combination of the age of the item in relation to the "Newer Than" and "Older Than" settings and the number of libraries in Group 2 and Group 3 that own the item.

Once *pWeight* is established, the formula to modify it will be determined. This formula is based upon whether Gift Manager or Deselection Manager mode is activated, and, if the Gift Manager is active, then whether the item is held locally. Based on those results, *pWeight* is modified differently by the Uniqueness and Collecting Level numbers. After the final *pWeight* is determined, a check is made to determine if *pWeight* is less than 1. If it is, then it is set to 1. The final weighting score is determined by dividing 10,000 by *pWeight* (see fig. 11 for the logic flowchart).

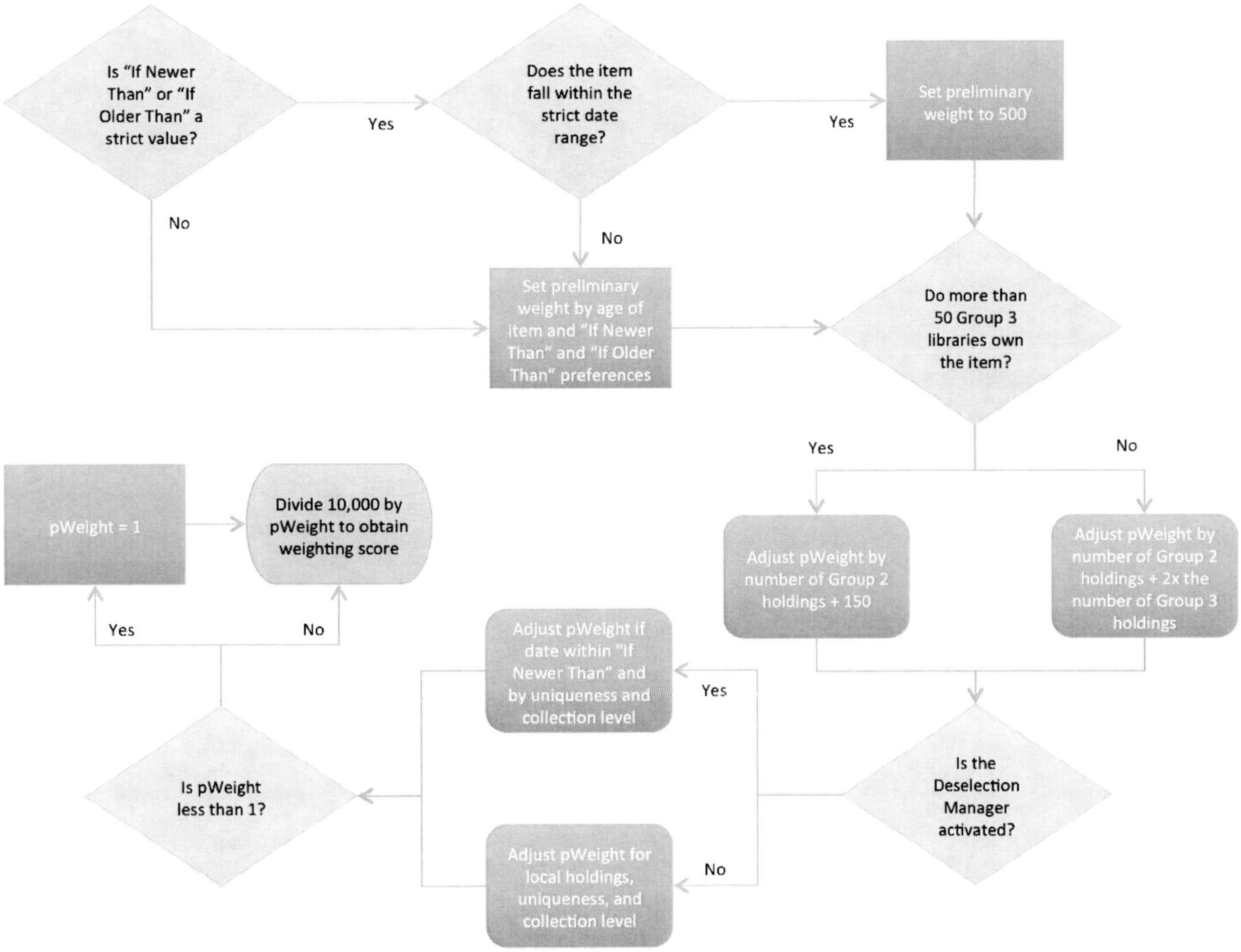

Figure 11. Logic for determining the weighting score in GIST GDM.

Chapter 8:

GIST GDM: Deselection and Collection Evaluation

By Kate Pitcher, Head of Technical Services & Collection Development, SUNY Geneseo

Deselection with the GIST Gift and Deselection Manager (GDM)

As mentioned in Chapter 7, libraries are grappling with increasing commitments and responsibilities coupled with (in many cases) an aging book collection, a decrease in use of book collection, and dwindling shelf space. There are several reasons for weeding the library's book collection, number one being the removal of outdated and obsolete material. No library is large enough to keep every book published, so we must make careful decisions, remove outdated books, and create a smaller, high quality collection that is relevant to the needs of current faculty and students. Figure 1 shows the enormity of the problem at SUNY Geneseo's Milne Library.

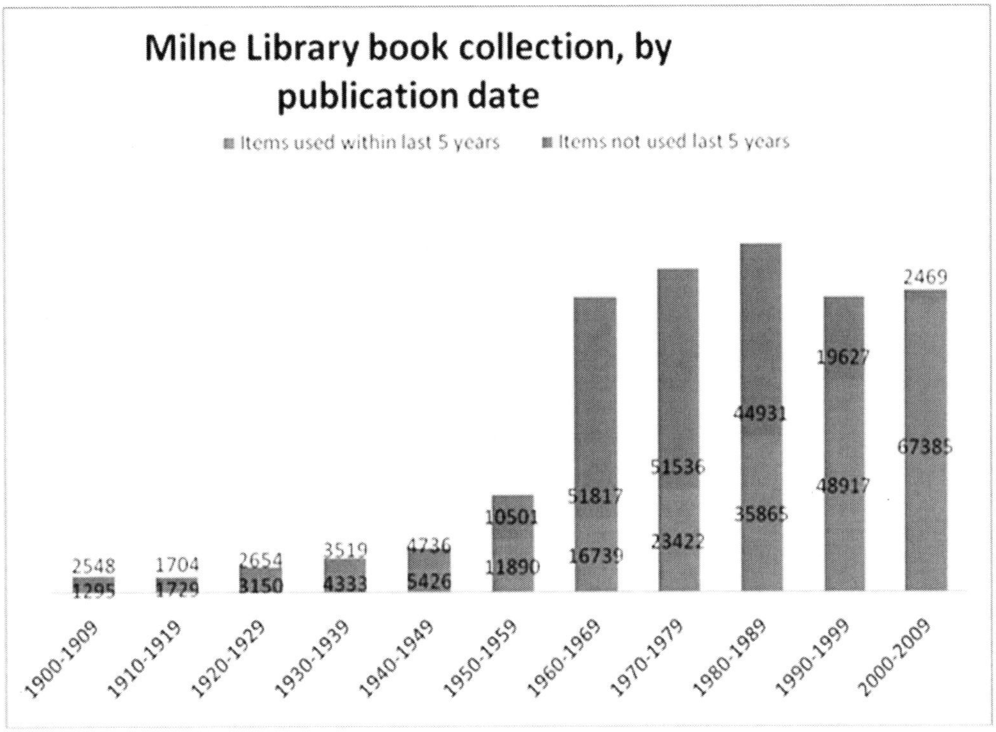

Figure 1. The Milne Library book collection by age of collection (publication date) and use of the collection (loans, renewals and in-house use).

While newer materials make up the majority of the circulated items, the majority of our collection is between 40 and 50 years old. While recent titles do circulate, over half of our collections are unused. This is a serious problem, but one that is not unique to Geneseo. In 2010, Cornell University Library's Collection Development Executive Committee Task Force on Print Collection Usage found that almost 45% of the print books in the Cornell collections published since 1990 circulated at least once, but that 55% of these books never circulated at all. The report also noted that "Circulation of monographs published since 1990 has tended to increase gradually for 12 years, at which point the use of new volumes tends to level off" (Rockey 2010). In addition, data from Milne Library interlibrary loan requests showed that students and faculty were requesting newer materials, published between 2000 and 2011. Sixty-one percent of our interlibrary loan requests between October 2009 and January 2011 were for items published in the last 10 years, indicating a need for newer materials in our collections. This leads to a decision point: we must deselect materials no longer useful or relevant to the users and programs at our institutions. In spring of 2011, Milne began an evaluation of the circulating collections, using ILS reports and the GIST GDM tool to provide data on usage and full-text availability information about our current print holdings: Was the item ever checked out and when? Are there trusted full-text free equivalents available online? How many copies are available in our Rochester region or SUNY libraries? Careful consideration of this data allowed librarians and faculty to analyze this information and consider deselection and other collection development considerations.

In the spring of 2011, Milne Library ran a zero usage report of the entire circulating collection and found that out of the 292,009 items in this collection at the time of the report, 199,520 did not circulate between 2005 and 2010. Sixty-eight percent of our circulating collection remained on the shelf during a five-year period. To improve the usage of the collection, begin building a newer and more up-to-date collection, and find room for the new services we want to provide, we needed to deselect the obsolete and we needed to do it with streamlined and data-driven decision making.

Item analysis

GIST GDM offers two ways to enable librarian or staff decision making for deselection: *Item Analysis* and *Batch Analysis.* When the deselection process is an item-by-item driven workflow, the default Deselection Manager interface (*Item Analysis*) works exactly like gifts processing, and applies the process to making decisions about weeding an item, rather than adding it to the collection (see fig. 2).

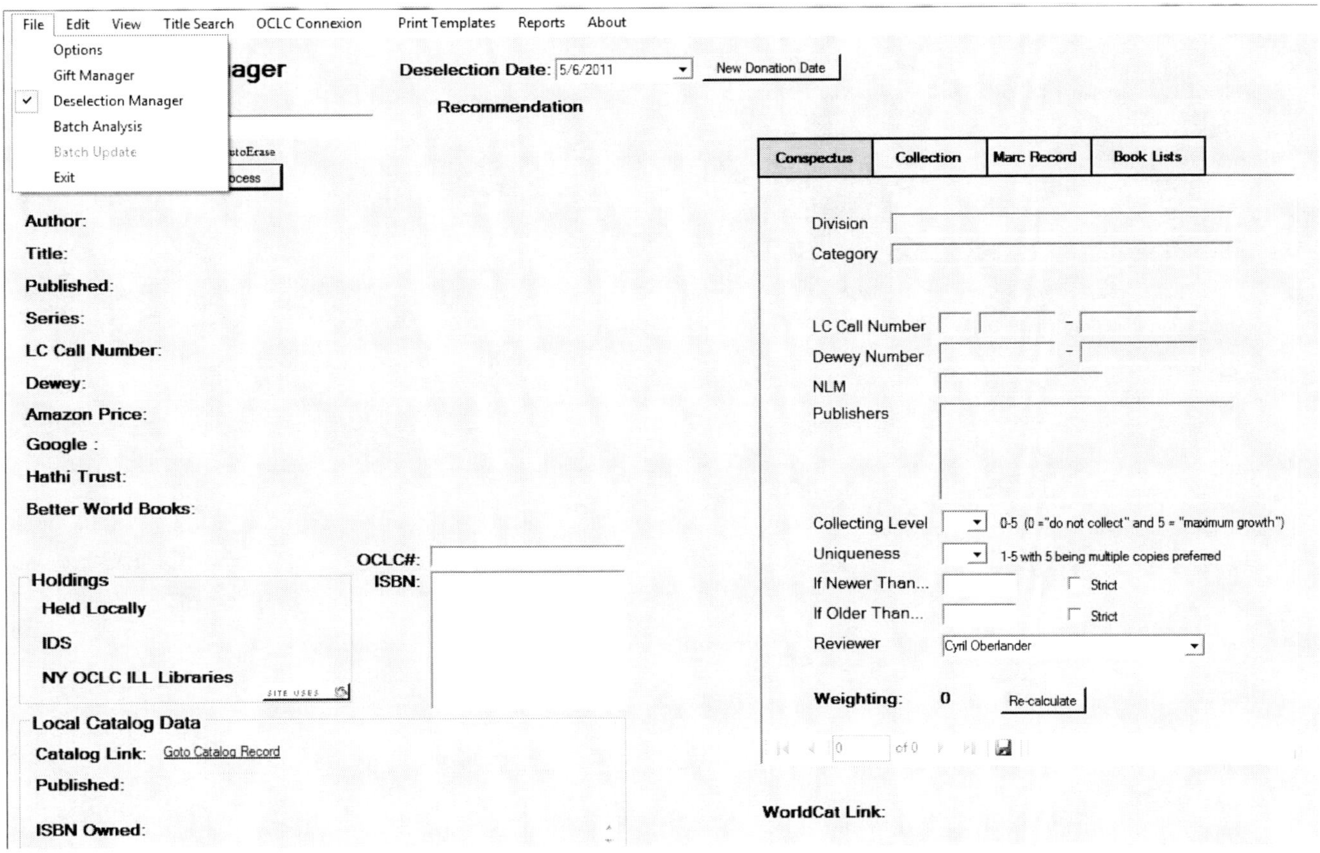

Figure 2. To do an item-by-item analysis for deselection, choose Deselection Manager from the File menu in the Menu bar.

When the Gift Manager is changed to Deselection Manager in the *File* menu on the taskbar, the interface looks the same. Because the item is in the collection, the weighting is adjusted. Wand in the ISBN, type in the OCLC number, or look up by title by selecting *Title Search* on the menu taskbar.

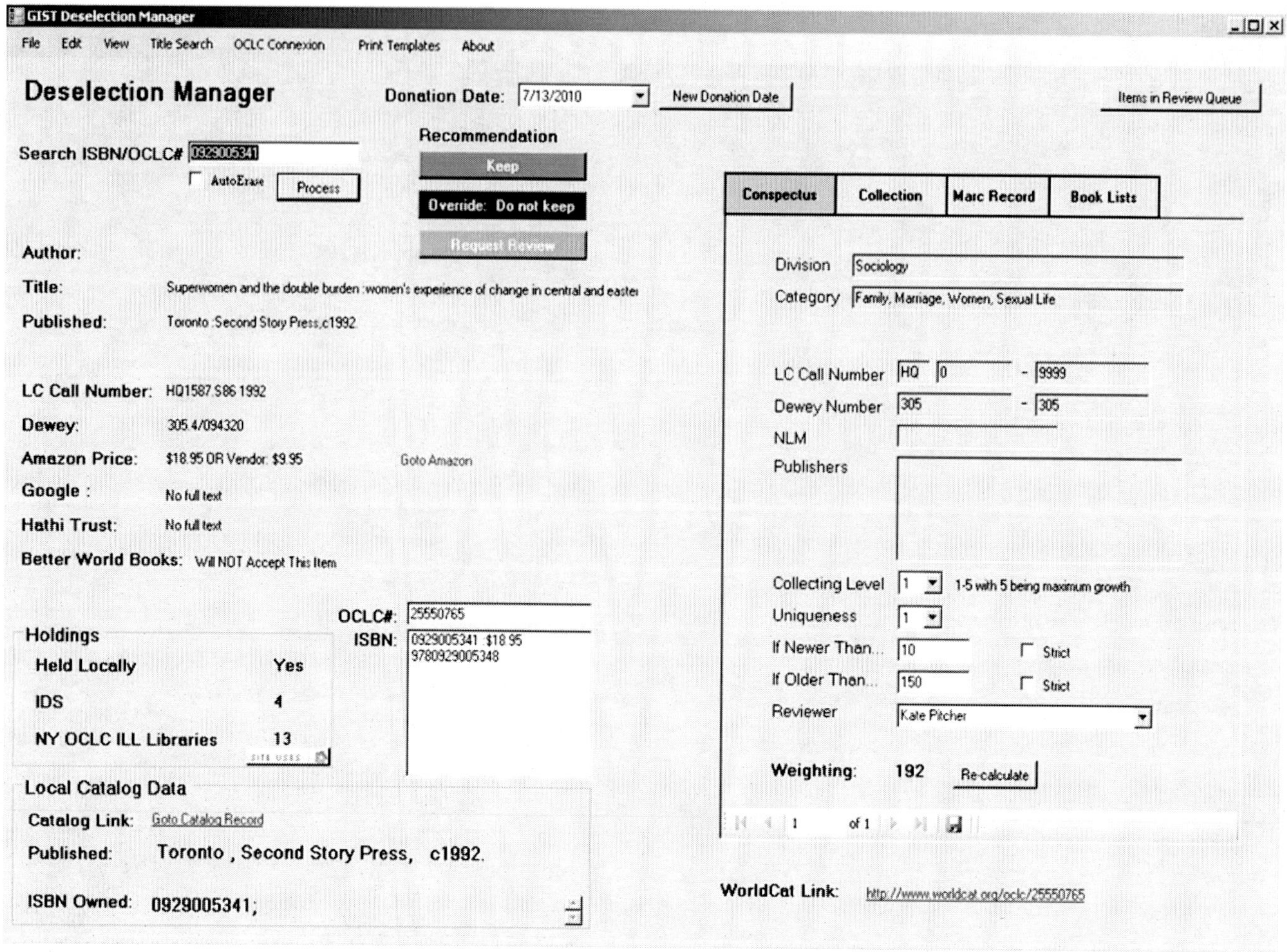

Figure 3. Results screen in Deselection Manager.

As in gift processing, Deselection Manager returns a *Keep* or *Do not Keep* recommendation based on your local and group holdings and collection building profile parameters (see fig. 3). If you decide to weed, you have immediate access to OCLC Connexion for removing your holdings from WorldCat by choosing *OCLC Connexion* from the menu taskbar (see fig. 4). This allows you to immediately remove your holdings from the record in WorldCat. This does not mean your bibliographic, holdings, or item records are removed from your local catalog—please see your cataloger or ILS systems administrator for information on how to remove your local records.

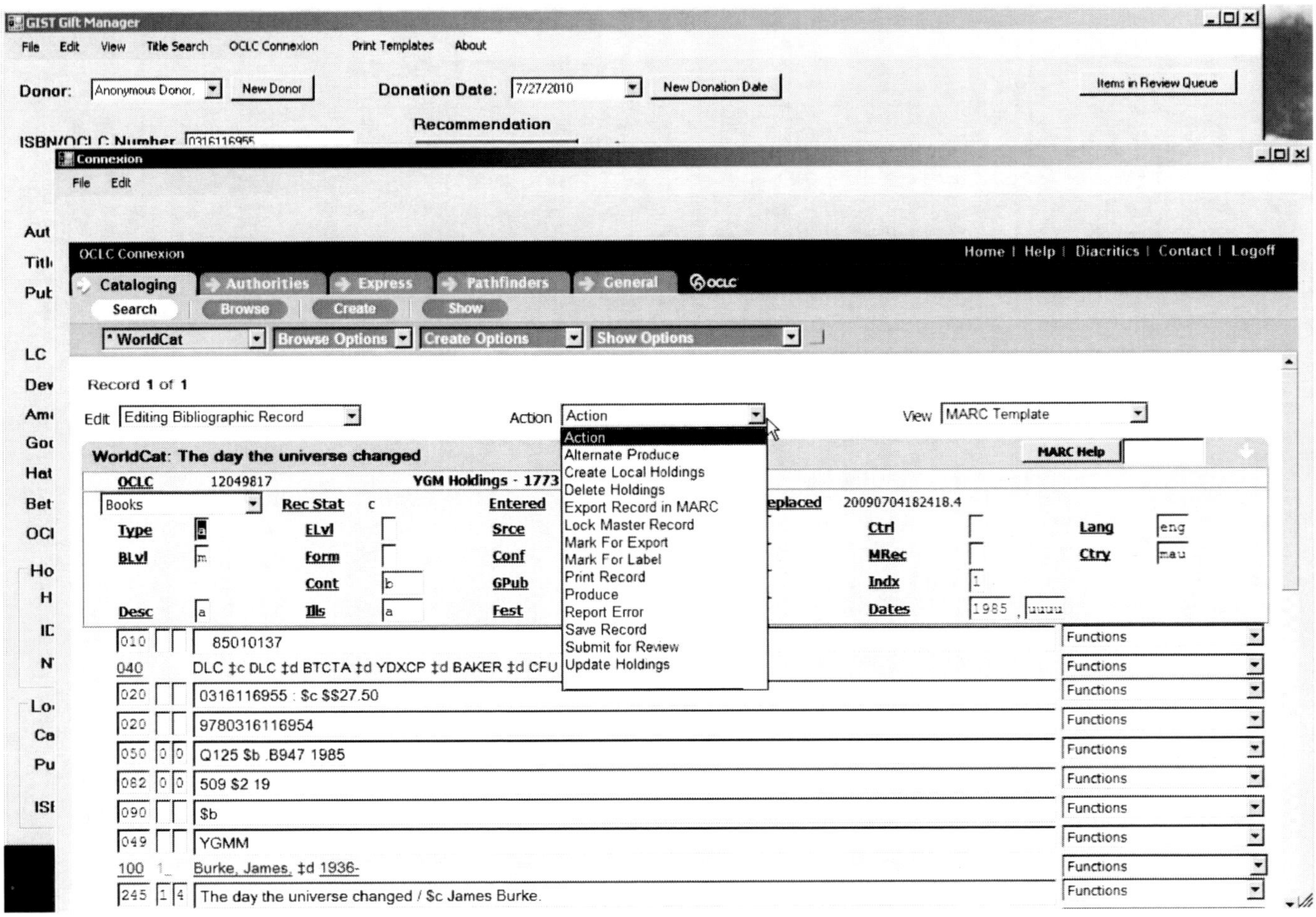

Figure 4. OCLC Connexion accessed from the Deselection Manager interface.

Batch analysis

For large weeding projects, GIST GDM also includes a *Batch Analysis* tool for providing useful evaluation information in a downloaded spreadsheet of multiple items. Library staff create a report of OCLC or ISBN #s from their ILS (for example, a zero use report of items never circulated), which then may be run through the *Batch Analysis* tool for further analysis (see fig. 5).

G6 fx 30260000670160

Bib Num	OCLC Num	Title	Author	Imprint	Year	Barcode	SubLib	CollCode	Call Num	Item Status
26594	1331956	... The Apology, Phaedo,	Plato.	New York, P. F. Collie	1909	30260000669824	GENMN	MCIRC	AC1 .A4 v.	
26602	1635844	... Essays and English trai	Emerson, Ralph Waldo, 1	New York, P. F. Collie	1909	30260000670111	GENMN	MCIRC	AC1 .A4 v.	
26606	1399226	...The poems and songs c	Burns, Robert, 1759-1796	New York, P.F. Collier	1909	30260000670137	GENMN	MCIRC	AC1 .A4 v.	
26613	1087137	Nine Greek dramas by Ae	Aeschylus.	New York, P. F. Collie	1909	30260000670152	GENMN	MCIRC	AC1 .A4 v.	
26615	18575559	...Letters of Marcus Tulli	Cicero, Marcus Tullius.	New York, P.F. Collier	1909	30260000670160	GENMN	MCIRC	AC1 .A4 v.	
26616	1835423	An inquiry into the natur	Smith, Adam, 1723-1790.	New York, P.F. Collier	1909	30260000669725	GENMN	MCIRC	AC1 .A4 v.	
26622	4247445	... The first part of the de	Cervantes Saavedra, Mig	New York P. F. Collier	1909	30260000669766	GENMN	MCIRC	AC1 .A4 v.	
26624	1337200	... Stories from Thousand	Arabian nights.	New York : P. F. Collie	1909	30260000669782	GENMN	MCIRC	AC1 .A4 v.	
26625	1492616	Folk-lore and fable: Aesop, Grimm, Andersen, wit		New York, P. F. Collie	1909	30260000669790	GENMN	MCIRC	AC1 .A4 v.	
26629	1474145	I promessi sposi (The bet	Manzoni, Alessandro, 17	New York, P. F. Collie	1909	30260000669840	GENMN	MCIRC	AC1 .A4 v.	
26631	1654933	...The Odyssey of Homer	Homer.	New York, P. F. Collie	1909	30260000669857	GENMN	MCIRC	AC1 .A4 v.	
26632	1455848	Two years before the ma	Dana, Richard Henry, 181	New York, P. F. Collie	1909	30260000669865	GENMN	MCIRC	AC1 .A4 v.	
26633	16729697	On taste : On the sublim	Burke, Edmund, 1729-17	New York : Collier., 1	1909	30260000669873	GENMN	MCIRC	AC1 .A4 v.	
26635	1210555	...English essays from Sir Phillip Sidney to Macaula		New York P.F. Collier	1910	30260000669907	GENMN	MCIRC	AC1 .A4 v.	
26636	1743521	...Essays, English and American, with introduction		New York, P.F. Collier	1910	30260000669915	GENMN	MCIRC	AC1 .A4 v.	
26640	721220	Scientific papers: physics, chemistry, astronomy, g		New York, P. F. Collie	1910	30260000669949	GENMN	MCIRC	AC1 .A4 v.	
26641	10672550	The autobiography of Be	Cellini, Benvenuto, 1500-	New York, P. F. Collie	1910	30260000669956	GENMN	MCIRC	AC1 .A4 v.	
26643	1692180	French and English philosophers: Descartes, Rouss		New York, P.F. Collier	1910	30260000669972	GENMN	MCIRC	AC1 .A4 v.	
26644	4199978	Chronicle and romance: Froissart, Malory, Holinsh		New York, P. F. Collie	1910	30260000669980	GENMN	MCIRC	AC1 .A4 v.	

Figure 5. Example of an Excel spreadsheet output from Geneseo's Aleph ILS system, reporting zero use titles.

Next, pull either ISBN or OCLC numbers from your spreadsheet and create a text file (using Notepad or another text editor) of just those numbers (see fig. 6).

Circulation1OCLCnumbers_MASTER - Notepad

File Edit Format View Help

```
1331956
1635844
1399226
1087137
18575559
1835423
4247445
1628927
1337200
1492616
1474145
1654933
1455848
16729697
1210555
1743521
721220
10672550
1692180
4199978
1837110
10680702
1199647
16771063
16738929
1837166
1393996
1538409
1990756
7480551
830444
369135
910651
910651
910480
912317
79326058
912349
912805
2519788
```

Figure 6. OCLC numbers in a text file output for GIST GDM.

From the main screen of GDM, go to the *File* menu and select *Batch Analysis* from the dropdown menu (see fig. 7).

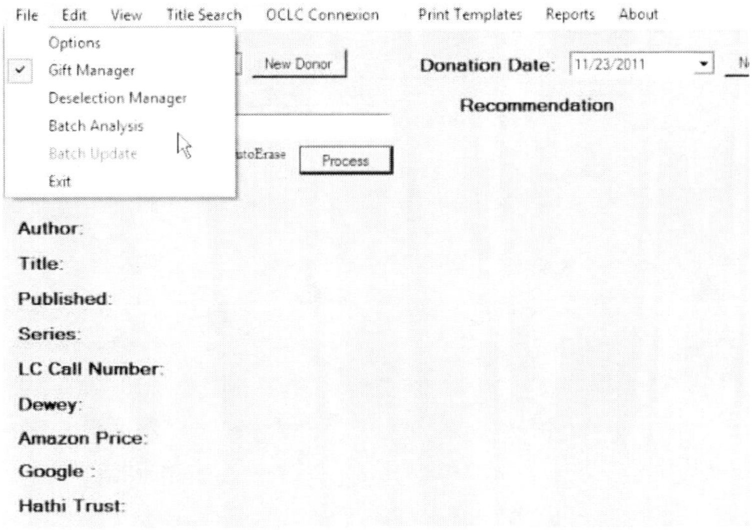

Figure 7. Select Batch Analysis from the File menu on the Menu bar to run an analysis of the ILS output.

In the Batch Analysis tool, select the .txt file you wish to run and then select the customized variables for your report:

- *Use Strict Dates for Conspectus*: automatically recommends "Do not Keep" for items outside your desired date ranges.
- *Ignore Uniqueness*: will not include your stated uniqueness preferences in the weighted recommendation.
- *Extreme Weeding*: forces the GDM to only keep items strictly within your conspectus criteria.
- *Add Records to Deselection Collection*: adds items in your batch analysis to a collection for deselected materials in your GIST GDM collections database.

A limited number of records may be processed by GDM in one day. The Google Books API has a 1,000 per day hit limit (more if you obtain a key), and the WorldCat API depends on what record limit you have asked for and received. Click on *Begin Batch Process* and watch as the records are processed (see fig. 8).

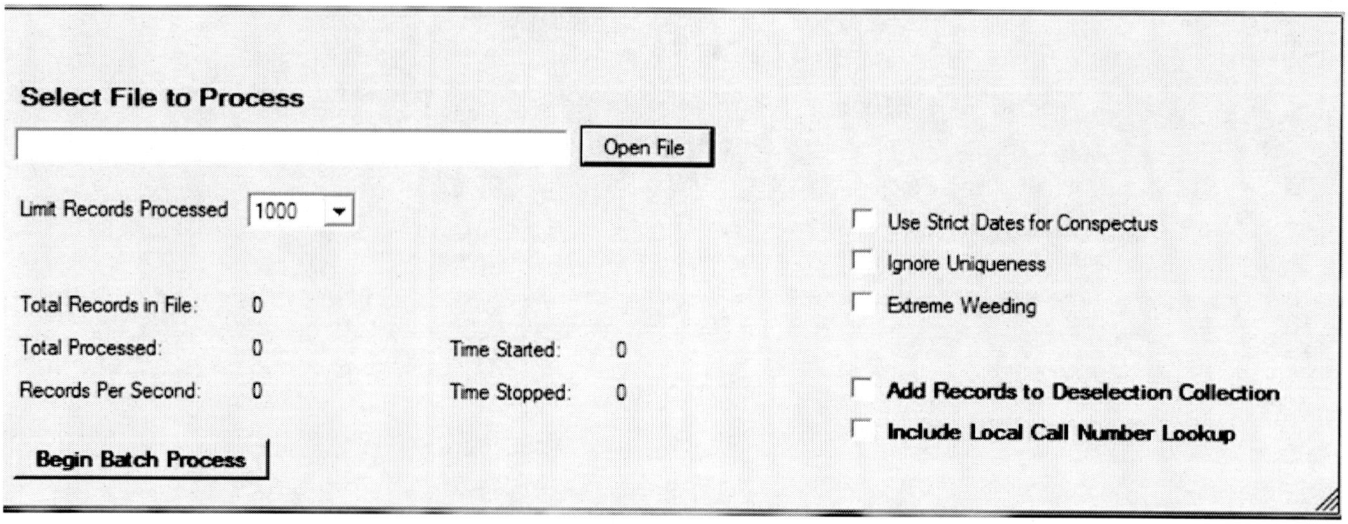

Figure 8. Batch Analysis processing.

When processing is finished, an Excel spreadsheet will open that can be saved to your computer (see fig. 9). Subject librarians and collection development staff can now review a report that includes the following data:

- Title, Author, Publication Date
- OCLC Call Number (050 or 090 field)
- Number of holdings in your groups
- Full Text in Hathi Trust and Google Books
- Better World Books library discard acceptance
- Weighting and conspectus recommendation for keeping or deselecting

	A	B	C	D	E	F	G	H	I	J	K	L	M	N	O
1	Title	Author	Original Search Num	OCLC Number	# of IDS	# of NY OCLC ILL I	Call Number	Local C	Pub Dat	Price	Hathi Trust	Google Books	Better Wor	Weight	Keep?
2	Viator.		1769080	1769080	16	58	CB3.V53|D2.V5		1997	No ISBN	No full text	No Full Text	No ISBN	☑	
3	Viator.		1769080	1769080	16	58	CB3.V53|D2.V5		1997	No ISBN	No full text	No Full Text	No ISBN	☑	
4	Viator.		1769080	1769080	16	58	CB3.V53|D2.V5		1997	No ISBN	No full text	No Full Text	No ISBN	☑	
5	Viator.		1769080	1769080	16	58	CB3.V53|D2.V5		1997	No ISBN	No full text	No Full Text	No ISBN	☑	
6	Viator.		1769080	1769080	16	58	CB3.V53|D2.V5		1997	No ISBN	No full text	No Full Text	No ISBN	☑	
7	Viator.		1769080	1769080	16	58	CB3.V53|D2.V5		1997	No ISBN	No full text	No Full Text	No ISBN	☑	
8	Viator.		1769080	1769080	16	58	CB3.V53|D2.V5		1997	No ISBN	No full text	No Full Text	No ISBN	☑	
9	Viator.		1769080	1769080	16	58	CB3.V53|D2.V5		1997	No ISBN	No full text	No Full Text	No ISBN	☑	
10	Viator.		1769080	1769080	16	58	CB3.V53|D2.V5		1997	No ISBN	No full text	No Full Text	No ISBN	☑	
11	Viator.		1769080	1769080	16	58	CB3.V53|D2.V5		1997	No ISBN	No full text	No Full Text	No ISBN	☑	
12	Viator.		1769080	1769080	16	58	CB3.V53|D2.V5		1997	No ISBN	No full text	No Full Text	No ISBN	☑	
13	Mediaeval studies.		2247258	2247258	14	56	D111.M44|CB351.F39			No ISBN	No full text	No Full Text	No ISBN	☑	
14	Mediaeval studies.		2247258	2247258	14	56	D111.M44|CB351.F39			No ISBN	No full text	No Full Text	No ISBN	☑	

Figure 9. Results of batch analysis.

Additional uses for Batch Analysis tool

GDM batch analysis has been used for JSTOR analysis, and can be used to identify items to move from open stacks to special collections, or to target local holdings that have not been digitized.

GDM is useful whenever a large number of holdings need to be analyzed:

- Compare duplicate periodical holdings in a consortia or group of libraries
- Identify local textbook holdings
- Identify unique collections in your library
- Identify materials to be digitized or scanned for publishing reprints
- Identify subject holdings across a region or consortia for cooperative collection development

One test of the *Batch Analysis* tool was to compare duplication of JSTOR titles across Ohio libraries in OHIOLink. OCLC #s provided by one Ohio library returned results across all the libraries in the OHIOLink consortium and indicated areas of diversity as well as duplication in the system.

At Geneseo, GDM helped us weed a large, unused print monograph collection in storage (31,000+ items). Using *Batch Analysis,* we found 13% of the collection was available full-text from the Hathi Trust repository of digitized books. In addition, we used GDM to do a batch analysis of our local history collection to determine which titles were rare, special, or unique to Geneseo and might be considered good candidates for reprinting and publishing using the Amazon CreateSpace publishing platform. To date, we have reprinted and published 33 titles in the Genesee Valley Historical Reprint Collection series, initially identified by using the GIST GDM *Batch Analysis* tool.

At a time when monograph budgets are declining, libraries can streamline gift processing, develop marketing techniques, and promote literacy campaigns by partnering with Better World Books. From July 1–November 1, 2010, Geneseo processed 3,000 items using GIST GDM, added 360 items to the collection (12%), and sent 2,640 items (88%) to Better World Books. Of the 3,000 items donated, 590 of these items (20%) were published between 2000 and 2010, a much higher figure than initially forecast when predicting what types of gifts we would receive. With GIST GDM, libraries can build current, topical, and relevant collections, streamline operations, and help literacy campaigns—all at the same time.

Resources

Globally changing records in ILS for a deselection project

The instructions for global changes are specific to each integrated library management system. Each ILS should have instructions for how to globally change, delete, or suppress bibliographic holdings or item records. See your local ILS systems or technical services librarian, or IT support desk for assistance.

Batch Deleting in OCLC

Start with a file of OCLC numbers for the items you are interested in analyzing. The easiest way is to get an output file from your ILS which includes the OCLC numbers in a report field. Next, you will need to remove the prefix (OCoLC) from the numbers. The easiest method is to do a "Find/Replace to remove (OCoLC)" and then "Replace All" (see fig. 10).

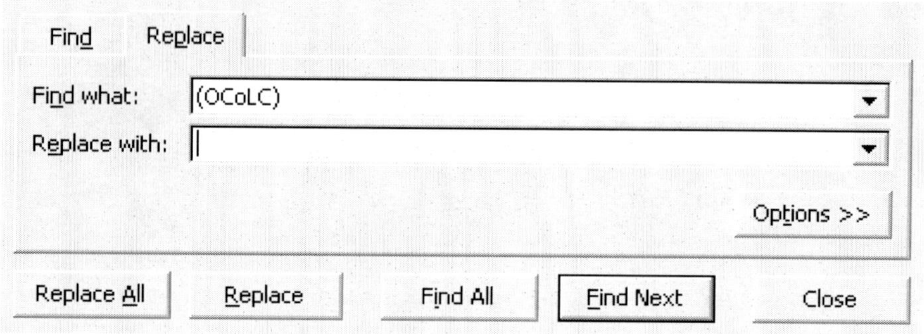

Figure 10. Use "Find/Replace All" to remove (OCoLC) from your OCLC numbers.

Save this new file of OCLC numbers as a text file in Notepad or another text editor and then log onto your Connexion client. Go to *Batch/Holdings by OCLC number* (see fig. 11).

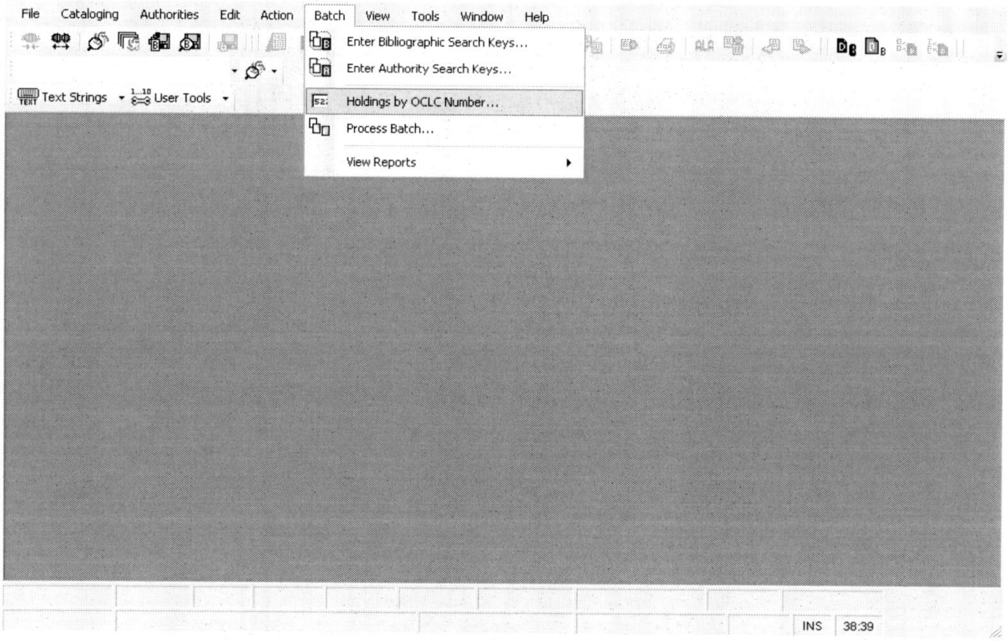

Figure 11. Choose Holdings by OCLC Number from Batch on menu taskbar in Connexion.

Next you will import your recently saved file of OCLC numbers into the client. Click *Browse* to find your file of OCLC numbers and then click *Import* (see fig. 12).

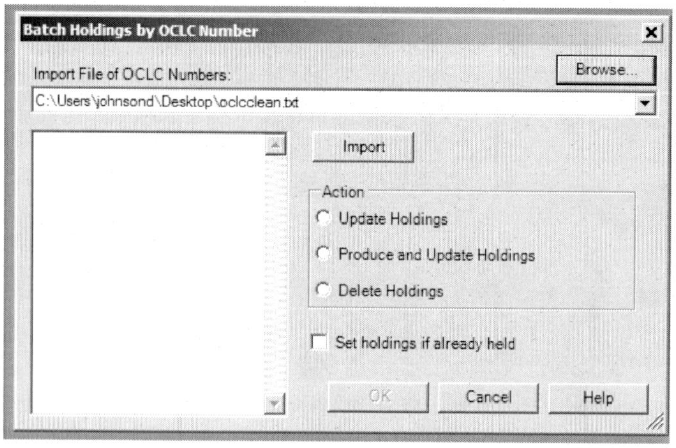

Figure 12. Import your files of OCLC numbers.

Next, a prompt for *Do you want to delete your original import file?* will appear. This is optional. If you have already saved your original file of OCLC numbers to your local computer, then click *No*. Otherwise, click *Yes* and follow instructions to save the import file to your computer (see fig. 13).

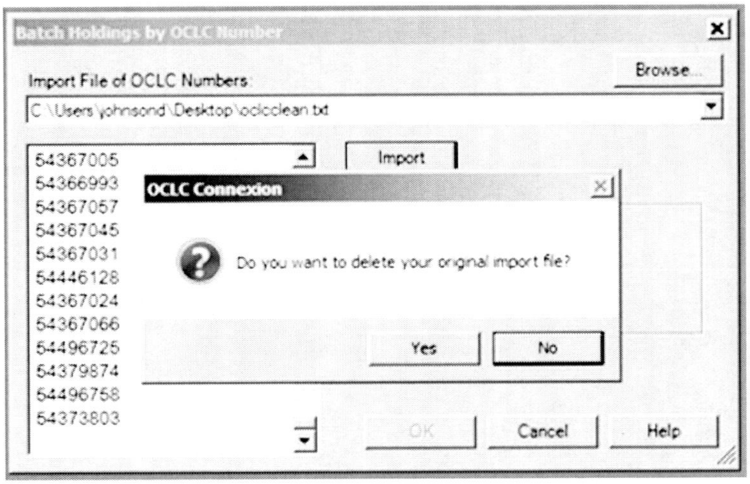

Figure 13. Prompt for deleting imported file.

Select the radio button for *Delete Holdings* and click *Ok* (see fig. 14). Evaluate the *Batch Holdings by OCLC Number Report* which appears and check for any problems or inconsistencies with the results.

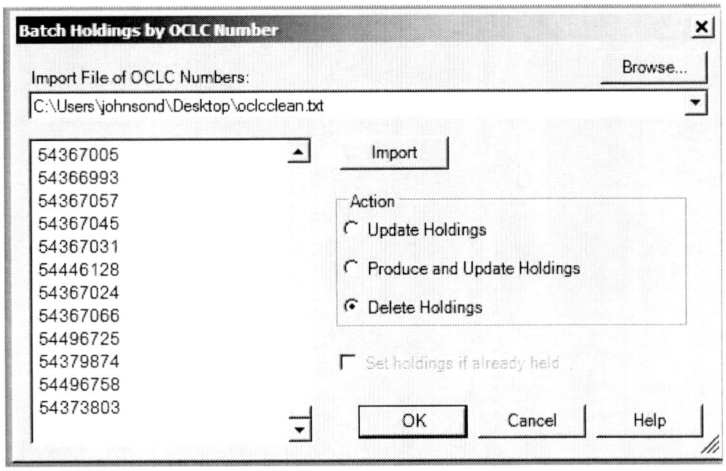

Figure 14. Delete Holdings screen.

Works cited

Rockey, S.W.,et al. "Report of the Collection Development Executive Committee Task Force on Print Collection Usage" *Cornell University Library*, 2010 http://vivo.cornell.edu/display/n5963.

Chapter 9:

Greasing the Book Truck Wheels: GIST at Washington University Libraries

By Eric Joslin and Barbara Rehkop, Washington University Libraries

Washington University Libraries first adopted the GIST Gift and Deselection Manager software in September of 2011, initially as an experiment to streamline the workflow of processing gift books. The GDM was installed on several staff machines in numerous departments using a network server. As we conducted testing, we quickly realized that the GDM in conjunction with ILS reports could be very useful in improving procedures beyond processing donations. Today, the GDM is used by multiple departments, including Access Services, Acquisitions, and our remote storage facility. In addition to using the GDM to improve our gift handling processes, we have also incorporated the software into our current book replacement procedures as well as in large scale deselection projects. Furthermore, our Acquisitions and Interlibrary Loan departments have incorporated the GIST Acquisitions Manager into their workflows in a buy-not-borrow pilot project.

Prior to installing GIST, a very large backlog of gift books had accumulated over several years. At that time student workers manually compared each volume to our catalog; books were set aside for subject librarians' review and eventually those not chosen were offered by the Library in a book sale. At the end of the sale, we were still trying to find other local groups willing to accept the remaining books, or recycling them.

Following the installation of GIST, students were able to identify duplicates much more quickly and accurately and provide additional information to the subject librarians about the holdings of other consortial partners, which had not been possible before. This concurrently decreased the time the librarians needed to make evaluations. We also discontinued our book sale and opted to send the duplicates and unwanted books to Better World Books. GIST improved this process by increasing speed and accuracy.

Now that the backlog is nearly gone, we are experimenting with sending librarians lists, arranged by donor, created in the Access database and exported into an Excel worksheet. Our hope is to further reduce the time the librarians must take to review gift books, and this looks promising. It would be helpful, though, if the information about our consortial partners were captured in the Access database.

The Batch Analysis function of the GDM has been particularly helpful in the revision of procedures for replacing missing books. In the past, the Circulation department used the ILS to generate a list of missing items based on a specific status field in the item record, and export the list to a spreadsheet. Once items were searched in the stacks, the list of unfound items would be forwarded to the Acquisitions department where it was posted on a Microsoft SharePoint site for replacement consideration by our subject librarians. At the time, only basic item record and circulation statistics were included.

GDM's capacity to configure consortial groups gave us the ability to incorporate far more information in the replacement lists. As part of the initial installation and configuration of the GDM, we assigned groups 2 and 3 of OCLC symbols to the Greater Western Library Alliance and MOBIUS, our most extensively utilized consortial partners. Now, once a replacement list from the ILS is produced by Circulation, the OCLC number is also exported as one of the fields in the spreadsheet. Acquisitions staff can then copy the OCLC number column from that list and save as a text file that can be opened and processed by the Batch Analysis function. The result gives us far more information on our missing items. With a simple copy and paste, Acquisitions staff is able to patch together a list of replacement considerations that combines ILS data with the output of the GDM report. The final product now includes basic bibliographic information, total circulations and renewals, item record notes, the number of GWLA and MOBIUS libraries that own an item, and static URLs to items that are available full text electronically in the Hathi Trust Library, of which we are members, or via Google Books.

With the new information we now incorporate into lists of missing books, subject specialists do not need to spend nearly as much time searching several catalogs to determine the availability of an item. In the past, replacement lists were posted irregularly and for months at a time. Now we have a much quicker turnaround time on review of the lists, which expedites both replacement time when necessary, and catalog cleanup for items we withdraw. The ability to see availability from multiple sources at once also prompted new replacement policies, wherein we no longer replace a missing item if it is available full text electronically, or if a specified number of our consortial partners hold an item.

Currently, we are using the Batch Analysis tool in conjunction with a large scale deduplication project. After identifying some 20,000 titles duplicated in our main library, it was determined that the majority of duplicate copies would be deselected. While we did not use the GDM's consortial searching capability to guide retention decisions for this project, we have been using it to identify titles that are accepted by Better World Books. Lists of duplicate titles for each of the four floors of our main library are loaded as spreadsheets onto tablets. As student workers pull duplicate items, they are scanned in our ILS with a specific count use designation in the item record used for weeding projects. A report is then created of items with the specific count use statistic and it is compared to the original list of duplicate items to troubleshoot any catalog errors such as incorrect barcodes or copy numbers. OCLC numbers are exported from the ILS report as a text file and run through the GDM's Batch Analysis tool. In this case, not as much information is needed as with the Acquisitions replacement list. The final report from the GDM is filtered to include only books that are accepted by Better World Books. The report is then pared down to title, author, call number, and publication date, and books are pulled and packed from a staging area. To date, approximately 5,600 books have been deselected; one out of eight is sent to Better World Books.

While the process of deselection is relatively new to our library, driven by increasing concern surrounding physical collections space, it has prompted discussion on future uses of the GDM as a weeding tool. For instance, our remote storage facility holds a large collection of books that are cataloged in Dewey Decimal, and have never been updated to the Library of Congress designation used for our general collection. Many of these titles are believed to be out of copyright. By creating an ILS report that focuses on the MARC 260 field, we hope to identify works published before 1923 and use the Batch Analysis tool to retrieve static URLs to full text versions in HathiTrust or Google Books. Print copies can then be withdrawn and electronic item records can take their place. This method may not yield results as effective as our replacement lists, since it relies on the existence of an OCLC or ISBN number in the item record.

Another aspect of the GDM we have not explored but would like to consider utilizing is the customizable conspectus. We envision that this tool could help us be even more precise when selecting gift book offerings, for instance, by delivering more information about our current and desired collection strengths.

The bulk of our GIST usage involves the GDM, but we have also incorporated the GIST Acquisitions Manager into a piece of our interlibrary loan processing. In the summer of 2011, a buy-not-borrow (BNB) pilot was proposed both to explore a closer partnership between the Interlibrary Loan and Acquisitions departments and to test different kinds of patron driven acquisitions. To be considered for purchase as part of this pilot, items had to be monographs, not owned locally or by reciprocals, unavailable electronically, less than $100, that could be ordered and delivered within two weeks. Initially, the ILL and Acquisitions staff used an email workflow to communicate orders. ILL staff continued to process items for the pilot using a set of custom queues within ILLiad: BNB Request Processing, BNB On Order, BNB Unfilled, and BNB Received. Acquisitions staff continued to place orders per their standard procedures once they received an email from the ILL department. This email workflow did cause confusion; in some cases, duplicate copies were purchased, as Acquisitions staff did not have access to the originally requested item in ILLiad. ILL staff also sent order requests to Acquisitions for items available electronically.

The first step to amending the communication issues was to add Acquisitions staff as ILLiad users. This allowed all involved staff to more accurately track a request by having access to the custom BNB queues. Secondly, we installed the GIST Acquisitions manager. By customizing the GIST Purchase Addon, we were able to select our preferred vendors and funds. In our case, we limited the vendors to GOBI, Alibris, and Amazon. Now when Acquisitions staff is processing an item, they are able to click on the Purchase Addon tab and see at a glance if the price range and availability warrant ordering as part of the BNB program. These tools, paired with the HathiTrust and Google Books Addons allowed us to expedite the order process itself, and avoid unnecessary purchase of books available electronically. The GIST Purchase Addon has also acclimated Acquisitions staff to using ILLiad daily. With the aid of the Purchase Addon in tightening the workflow between departments and speeding up the ordering process, the BNB pilot has largely been a success. Of the titles we purchased, 25% were checked out beyond the original use, and 44% of those titles were renewed.

Since adopting and customizing GIST software, Washington University Libraries has significantly improved gift book processing and expedited the task of replacing missing books. The Gift and Deselection manager has proven to be a valuable asset in deselection projects, and the Purchase Addon has streamlined a cooperative effort between Acquisitions and Interlibrary Loan. While GIST's offerings are not yet widely used throughout all departments of our organization, we can see its potential for future weeding projects, and for increasing subject librarian participation in both the gift and deselection processes.

Conclusion

By Mark Sullivan, IDS Project Executive Director, SUNY Geneseo

The Getting It System Toolkit began as a system that combined ILL and Acquisitions by empowering users and staff with useful data for decisions and streamlining workflows. It expanded to gifts and deselection because that same collection profile and data were useful and critical to streamlining staff-intensive work.

GIST will remain a two-pronged system supporting both ILLiad/Patron Driven Acquisitions and gift processing/collection analysis. The IDS Project has adopted GIST and will be providing continuing support for the two systems. IDS has long focused on a community of trust and support for resource sharing and hopes to expand that philosophy to acquisitions and collection development.

In October, 2011, Atlas Systems announced that they would provide training for all libraries interested in implementing GIST for ILLiad. In November, 2012, OCLC approved the installation of new GIST tables into the ILLiad databases for all hosted sites. The support of both Atlas and OCLC has made the improvements and expansion of GIST for ILLiad possible. The future of GIST for ILLiad is dependent upon the resource sharing and acquisitions communities working together and how ILLiad evolves to the changing information delivery landscape. New versions of the GIST for ILLiad software will be released periodically when bug fixes or enhancements are developed. Most of the current enhancements come from user requests, such as the ability to use Google Books instead of Amazon for book descriptions and cover art.

The IDS Project has made a dramatic shift in the way the Gift and Deselection Manager works and GDM will be going through a major redesign during 2014. This online version of GDM will solve a myriad of problems that are caused by libraries using outdated operating systems and security profiles that limit installations and updates. Future benefits, however, would come not from the elimination of the local installation, but from the new functions that would allow for consortial-wide coordinated collection development. Coordinated collection development has been a goal to scale shared benefits of libraries for years and has yet to be successfully attained. Through CCD, acquisitions will become more cost effective by reducing expensive duplication of materials across several campuses. With resource sharing among libraries, student access to specialized materials would not be impeded. In fact, funding that would have been spent on duplication could be spent on alternative materials.

By linking gift processing, collection analysis, and deselection among consortial libraries, GDM Online would provide libraries with a way to juxtapose their data against other consortial holdings, thereby highlighting unique items and allowing for a reduction in duplication. This system would also allow for enhanced acquisitions through usage data from interlibrary loan systems, book lists, and circulation statistics. Book lists, for items such as textbooks, would allow libraries to determine if a gift received at one library was on a list at that library or at another member library. Reallocating textbook gifts to campuses in need would strengthen library textbook reserves and reduce the cost of learning for students, families, and campuses. Weeding of collections would also be possible on a consortial scale and would provide for a simple way to determine if one library's discards would fit another library's collection for areas in which they would like to grow. The last copy of a given title could be easily determined and would reduce the risk of libraries removing an item that would no longer be available within their consortium. GDM Online could also provide the option to coordinate the transfer of a title's last copy to a high-density remote storage facility. The IDS Project hopes to have GDM Online ready for trial in early 2015.

With the support of Atlas Systems and OCLC, GIST for ILLiad and the Gift and Deselection Manager will continue to be successful and to provide libraries with the ability to handle Patron Driven Acquisitions and to manage their collections more efficiently. With support from the community, both systems will be able to grow and improve as new ideas and processes are implemented.

Made in United States
North Haven, CT
05 November 2026

21772141R00050